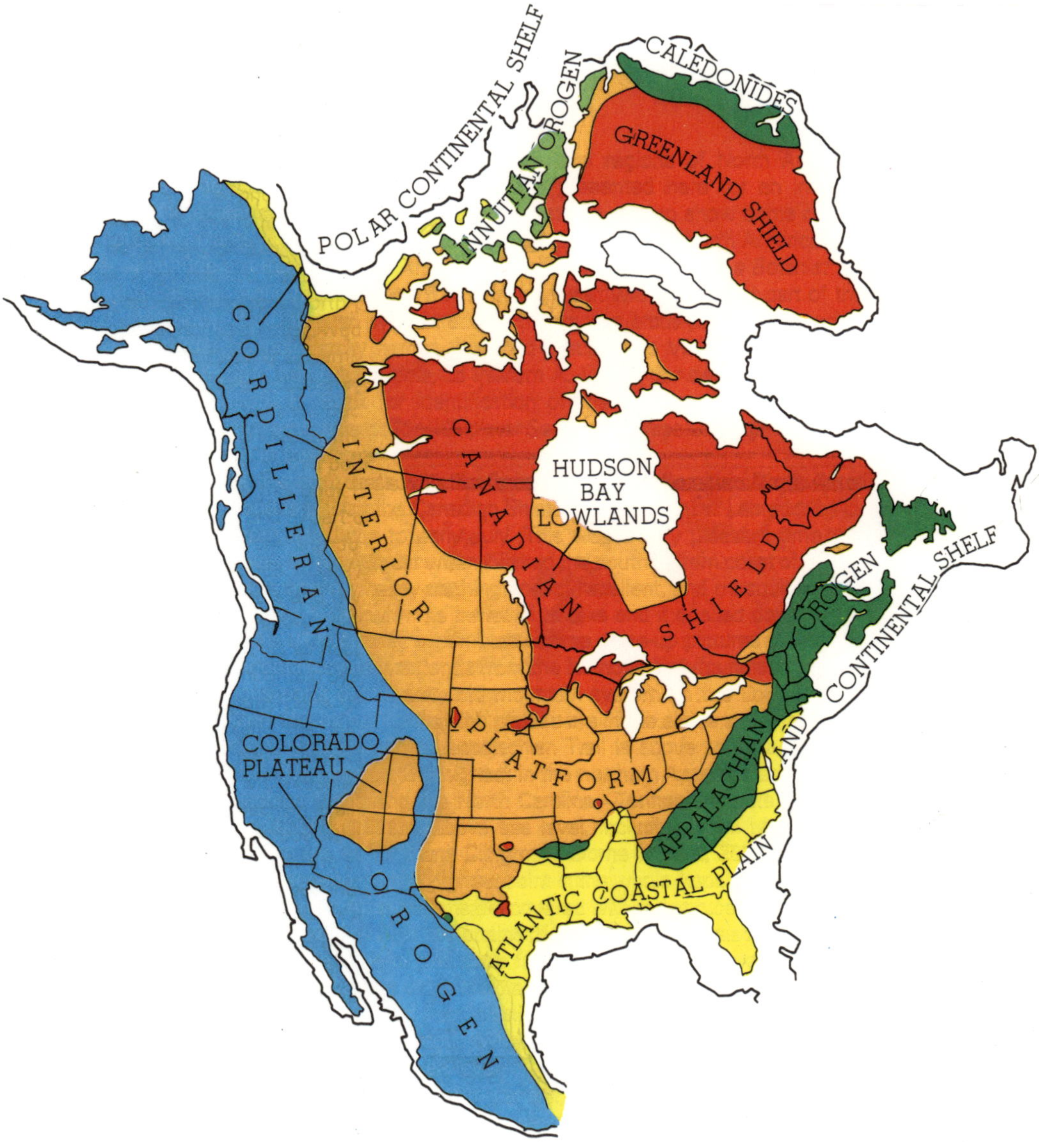

Major Structural Units of North America

Courtesy of the Department of Earth Sciences, Memorial University of Newfoundland

The central, broad, stable areas are the Shield of exposed Precambrian (ancient) rocks, and the Platform of relatively thin cover rocks (covered Shield). They are surrounded peripherally by belts of deformed rocks less than 600 million years old — the Appalachians, Cordillera, Innuitian, and Caledonides. These are followed outward by the modern continental margins of Mesozoic and Cenozoic mainly sedimentary rocks that form the modern coastal plains and continental shelves.

EASTERN NORTH AMERICA AS SEEN BY A BOTANIST: PICTORIAL

I. THE ARCTIC REGION

EASTERN NORTH AMERICA AS SEEN BY A BOTANIST: PICTORIAL

I. THE ARCTIC REGION

BY

In-Cho Chung

Printed by
Samhwa Printing Co., Ltd.
Seoul, Korea
1989

Available from
In-Cho Chung
1308 Laurel Drive
Daytona Beach, Florida 32017
U.S.A.

Contents

page

Preface and Acknowledgements 7
I. The Arctic Region 9
Common Arctic and Subarctic Plants 66
Selected References 106
Index 108
Appendix 111

II. The Wooded Regions 113
The Subarctic Canadian Shield Region 113
The Appalachian Region 133
The Canadian Appalachians 134
The U.S. Appalachians 183
New England Section 183
North Carolina — Tennessee Section 211
Virginia — West Virginia — Pennsylvania Section 256
The Coastal Plain Region 287
Common Mushrooms of the Temperate Region 369
The Families and Genera of Fungi in this Book 404
Selected Referencea for the Wooded Regions 405
Index for the Wooded Regions 412

The author (center) in the High Arctic, August 1987

In-Cho Chung was born in North Korea in 1918. He attended the University of Tokyo, Seoul National University, and later the University of Michigan where he received his M.S. and Ph.D. He was assistant curator at the Botany Department of Field Museum of Natural History in Chicago and taught at Daytona Beach, Florida; Chicago; and Mansfield, Pennsylvania as a professor of botany. Dr. Chung has traveled widely across the Canadian Northland and Alaska, through the Rocky and Appalachian mountains, and along the East Coast. His recent publications include *The Arctic and the Rockies as seen by a Botanist: Pictorial.*

Preface and Acknowledgements

During the past two decades I have visited various regions of North America for botanical observations. The photographs from the East are presented here for an interpretation of the landscapes, to examine the principal vegetational features, and to show some of the characteristic and interesting plants in these regions. Since the arctic regions of North America south of the 75th parallel of latitude are treated in *The Arctic and the Rockies as seen by a Botanist: Pictorial* (Chung 1984), the Eastern Arctic north of the 75°N is mainly dealt with in the first part of this book for the sake of continuity. It is followed by illustrations of common arctic plants arranged generally in the alphabetical order of their family names. I extend gratitude to William J. Cody for checking the identifications of some flowering plants, to William A. Weber for determinations of the Canadian lichens, and to William R. Buck for identification of the mosses. I am also indebted to Bezal Jesudason of the High Arctic International Explorer Services at Resolute Bay, who kindly loaned three photographs included in this book.

The second part of this book deals with the forested regions of eastern North America. Subarctic Labrador is wooded and is part of the Canadian Shield. I visited central Labrador by car from Goose Bay to near Esker from late June to early July. This was a very pleasant time to camp without mosquitoes or black flies. In late August when I visited the southeastern coast of Labrador there were millions of biting insects and my heavy clothing, insect repellent, and mosquito net to wear did not help too much. Fortunately I had with me benadryl to take and it quelled down my terribly swollen eyelids in five days. This all happened when I took some close-up pictures of lichens.

The Appalachian mountain system extends from the Island of Newfoundland to northern Alabama. I have sampled various parts of this very old mountain system. In the Long Range Mountains of western Newfoundland, the higher tops of the plateau are in the alpine zone. In the White Mountain National Forest of New Hampshire, the Appalachian Trail is above treeline for nine miles on the Presidential Range. Mt. Washington, the highest in the northern Appalachians, is 6288ft (1917m) in elevation. The Appalachian Trail along the North Carolina—Tennessee state line goes over several mountains higher than 6000ft (1829m) above sea level, including Roan Mountain famous for the Rhododendron Garden on the summit and Clingmans Dome (6643ft, 2025m) which is the highest point on the Trail and the second or third highest point in the whole Appalachians. The summits of the high mountains of the southern Appalachians are covered with red spruce and Fraser fir in contrast to the treeless alpine summits of the high mountains of the northern Appalachians. I am grateful to the rangers and naturalists of the national and state forests, parks, nature preserves, and Duke Power Company who assisted me in various ways in obtaining photographs of certain plants in bloom in this region. I also thank S.P. Vander Kloet of Acadia University in Wolfville, Nova Scotia, for an invitation to join the field trip to Brier Island, Nova Scotia, where *Geum peckii*, an endemic of the alpine areas of New Hampshire and Maine, grows at sea level.

The coastal plain region seouth of the 40th parallel, that is from New Jersey to Florida, was not directly affected by the last Ice Age unlike the northeastern United States and Canada. A long chain of barrier islands along the coast line protects the mainland and enclosed salt marshes from direct attack of the open ocean. Although I have never got lost in the Arctic, I learned that it is very easy to get lost in southern swamps. After I got in a rather dry swamp on one occasion I could not find my way out by the sunset. So I had to lie down on the ground in a dry spot, covered myself with litter, and waited until the day broke next morning when I saw quite a few rattle snakes around me. But I was satisfied with a few swamp pictures I took.

In Big Cypress Swamp of southern Florida I saw several orchids in fruit, but I could not get a picture of the flower of Night-smelling Epidendrum, because it opens during the night and does not last more than one day. I am grateful to C. A. Luer for loaning a picture of the orchid's flowers included in this book.

The common mushrooms of the temperate region in this book are based on my study of Pennsylvania mushrooms. I am greatly indebted to the late Alexander H. Smith, professor of fleshy fungi at the University of Michigan, who aroused my interest in this fascinating group of plants and did many favors including checking all but a few photographs of fungi in this book.

The geological maps on the endpapers are reproduced from those published by the Department of Earth Sciences, Memorial University of Newfoundland, with permission of Harold Williams who kindly assisted me in interpreting them.

My daughter, Linda, was very helpful in composing the photographs of the dwarf-scrub lands of northern Cape Breton Island and Newfoundland, and of the Plains in the Pine Barrens of New Jersey. My wife, Won-Yang, assisted me in photographing coonties and banana flowers in Florida.

I. The Arctic Region

Arctic lands are the areas beyond the polar treeline where the average July temperature is below 10°C (50°F). In eastern Canada, most of the arctic region lies north of the 60th parallel. Here the highest land in the Canadian Arctic occurs along the eastern coasts of Baffin, Devon, and Ellesmere islands and also in northern Ellesmere and Axel Heiberg islands. The rugged mountains ranging from 1500 to 2600 m (5,000 to 8,500 ft) high are composed of crystalline rocks on Baffin Island and of sedimentary rocks in northern Ellesmere Island. From south to north, average annual precipitation generally decreases from about 650mm (25.5 inches) on southeastern Baffin Island to about 65mm (2.5 inches) in the Eureka-Lake Hazen region of northern Ellesmere Island. This is the least mean annual precipitation in Canada and is the result of the rainshadow effects, the exposed slopes of surrounding mountains receiving more than 200mm. In the High Arctic, mostly north of Parry Channel (comprising the wide channels between M'Clure Strait around the 75th parallel in the west and Lancaster Sound in the east), July mean daily temperature is below 5°C (41°F). Winter arrives by the end of August in the High Arctic and by the end of September in most of the Low Arctic; and spring does not come until mid-June in most areas. It snows least (about 45 cm or 17.7 inches) in the Eureka-Lake Hazen region and most (up to 600 cm or 236 inches) in southeastern Baffin Island. The midwinter is very cold and continuously dark: for about 45days in the middle-arctic region around the 70th parallel; 90 days at Resolute Bay (74°43'N), 120 days at Eureka (80°0'N), and 135 days at Alert (82°30'N) in the High Arctic. During the two coldest months, January and February, temperatures average from about -25°C (-13°F) in the South to about -35° (-31°F) in the Far North adjacent to the Arctic Ocean. This is also the time when the sun first rises in the Arctic. Sooner or later most of the Arctic has no sunset. The sun is continuously above the horizon for approximately 150 days from early April to early September at Alert, 140 days at Eureka, 100 days from May to mid-August at Resolute Bay, and 70 days from mid-May to late July at 70°N. Arctic lands are also characterized by continuous permafrost. An unbroken expanse of perennialy frozen ground (except under deep lakes and rivers) underlies a thin surface layer of soil which thaws in summer and freezes in winter. Permafrost and intensive frost action affect the vegetation. The shallow, unstable, and nutrient deficient soils do not favor plant growth.

1. Sunrise in mid-February at Grise Fiord (76°25'N), Ellesmere Island. Photo by Bezal Jesudason 1971.

1

2-3. Low Arctic Iqaluit (Frobisher Bay), southern Baffin Island, 07-19-87. The Low Arctic includes the southernmost part of Baffin Island, most of the mainland areas of arctic Canada, and nearly all of the northern Alaska mainland north of the Brooks Range. The summer temperatures here are high enough for the formation of birch and willow thickets in suitable places. Most low-arctic areas are covered with continuous vegetation, including many dwarf shrubs such as Northern Labrador-tea (#4), Lapland Rosebay (#5), Alpine Bearberry (#195, 196), Alpine Azalea (#202), and Mountain Cranberry (#204, 205). These members of the heath family (Ericaceae) do not occur in the High Arctic. Other plants growing here include Maydell's Oxytrope (#6) and Alpine Milk-vetch (# 206) which are members of the pea family (Fabaceae, Leguminosae) not found in most high-arctic areas; Diapensia (#191), Black Crowberry (#192), Arctic Heather (#198), Alpine Bilberry (#203), Arctic Willow (#240, 244), and Moss Campion (#44) which are also found in the High Arctic. Lichens (#266, 271) fill the gaps between the flowering plants and rocks.

2

3

4

4. Northern or Narrow-leaved Labrador-tea, *Ledum decumbens* (*L. palustre* subsp. *decumbens*), Iqaluit, Baffin Island, 07-19-87. The flowers are similar to those in #201. The leaves are linear, with the margins tightly rolled under; and densely rusty woolly hairy on the undersides (seen at cross section). Young twigs are rusty woolly hairy. Fruit stalks are sharply bent near the ends.

5

5. Lapland Rosebay, *Rhododendron lapponicum*, Auyuittuq National Park, Baffin Island, 07-16-78. The leaves and capsules are densely covered with rusty-brown scales.

6

6. Maydell's Oxytrope, *Oxytropis maydelliana*, Pond Inlet, Baffin Island, 07-05-83. The base of the hairy plant is densely covered by chestnut-brown stipules of dead leaves.

7

7. Habitat of Moss Heather, Iqluit (Frobisher Bay), Baffin Island, 07-19-87. Moss Heather (#8-9) and Mcuntain Heather (#10-11) grow in the low-arctic, subarctic, and alpine areas where snow remains late in spring. They do not extend much beyond the Arctic Circle (66°30′N) on Baffin Island.

8

9

8-9. Moss Heather, *Cassiope hypnoides* (*Harrimanella hypnoides*), by a brook in #7, 07-11-85. The Moss Heather grows here with Horsetail (*Equisetum variegatum*), Snow Buttercup (#231), Herb-like Willow (#246, 247), and Nodding Saxifrage (#251). The Moss Heather is a small, prostrate plant with crowded needle-like leaves resembling a moss when without flowers or fruits. These small, nodding, bell-shaped flowers are about 5mm across. The fruits are small, subglobose capsules.

10 

11 

10-11. Mountain Heather, *Phyllodoce caerulea*. 10: Mt. Albert, Québec, 07-02-74; 11: Iqaluit, Baffin Island, 07-11-85. This low, much branched, evergreen shrub is often found associated with the Arctic Heather (#198) and Alpine Bilberry (#203) at Iqaluit. The crowded, linear, flattened leaves are rolled under, forming a groove, and the leaf margin is minutely toothed. The persistent, peg-like leaf-bases make the twigs rough. The calyx lobes and flower-stalks are densely glandular-hairy.

12 

12. Dwarf Birch, *Betula glandulosa*,, Iqaluit, Baffin Island, 07-17-87. The leaves of this birch are rounded at the apex, round-toothed on the margins, and more or less wedge-shaped at the base. The twigs are densely covered with wart-like resinous glands. It is a common low shrub up to head-high in the subarctic and subalpine areas. It reaches its northern limits of distribution in southern Baffin Island. Here (#12) it grows in isolated, prostrate patches in snow accumulation areas with Herb-like Willow.

13

14

13-14. High Arctic, Resolute Bay area, Cornwallis Island, 06-22 07-09-83. The dry, stony and rocky areas are nearly barren with scattered Purple Saxifrage (#253-255) and Arctic Willow. Better vegetation is found only in places below melting snow-drifts, around small lakes and ponds, and along streamlets. In #14, Arctic Willow and Arctic Avens (#233, 234) grow on the slope below the snow; in the lower wet place, Sulphur-colored Buttercup (#15), Spider Plant (#26), Cotton-grass (#37), Yellow Marsh Saxifrage (#38), Nodding Bladder Campion (#40), and Alpine Foxtail (#220) grow.

15

15. Sulphur-colored Buttercup, *Ranunculus sulphureus,* in #14, 07-09-83. The sepals are covered with dense, long, dark-reddish-brown hairs. See also #36.

16-18. High Arctic, Eureka, Ellesmere Island, 07-27-85. The vegetation of the high-arctic areas (the Far North) is generally poor and sparse, often extremely so.

16

16. A dry, lower slope of Black Top Ridge. The occasional Vahl's Cinquefoil (#91, 92) was the only prominent plant found here.

17

17. Cracked clayey ground. Only very widely scattered herbs grow here, including *Melandrium triflorum* (#18), *Braya thorild-wulffii* (#19), and Red-stemmed Cinquefoil (#70). The bird is a Long-tailed Jaeger.

18

18. *Melandrium triflorum* (*Silene sorensensis*), Caryophyllaceae (Pink Family), on dry clayey ground near the seashore. The flowers are often three together at the end of the stem. This is a mainly high-arctic species of Canada and Greenland with small, more or less triangular, wingless seeds. This densely hairy plant also grows in very dry, gravelly places (#111) together with Arctic Bladder Pod (#62) and Red-stemmed Cinquefoil (#70).

19

20

19-20. *Braya thorild-wulffii,* near the fiord shore, Eureka, 07-30-85. This is a very small, high-arctic species. The plants in #19 were only about 1-2cm tall. The clustered flowers have obtuse sepals with scarious margins. The flowering stems, ovaries, and the ellipsoid pods (5-8mm long) are densely hairy, and the leaves are moderately hairy on the margins. The fruiting stems are ascending to prostrate. Brassicaceae (Cruciferae), Mustard Family.

21

21. *Taraxacum pumilum,* on clayey ground near the fiord shore, Eureka, 07-30-85. This mainly high-arctic dandelion is also a very small species. The plant in the picture was only about 5cm (2 inches) tall. The spreading leaves are deeply and regularly lobed into almost equal spreading lobes.

22

22. Many other species are widely distributed throughout the Arctic. Such plants may be found in the mounds of mosses in a wet place by a stream in the High Arctic as here, Little Cornwallis Island, 07-17-85. Such species growing here include Two-flowered Rush (#23, 24), Alpine Cress (#25), Spider Plant(# 26, 27); Scurvey-grass (#27), Yellow Marcsh Saxifrage (#38), Alpine Foxtail (#220), and Nodding Saxifrage (#251). The mosses here are *Orthothecium chryseum* and *Cinclidium arcticum*.

23

24

23-24. Two-flowered Rush, *Juncus biglumis*. The plants in the pictures were about 5-10 cm (2-4 inches) tall. The two flowers and fruits are inserted one above the other and subtended by a long, dark brown bract overtopping them. The other bract is much smaller. The dark brown, club-shaped trigonal capsules are shallowly notched at the flattened apex. These loosely tufted plants have few leaves only at the base. Juncaceae, Rush Family, see p. 83.

25

25. Alpine Cress, *Cardamine bellidifolia, in #22, 07-17-85.* These tufted dwarf plants in a cushion of moss were only about 3 cm tall including the clustered, elongated, nearly terete or more or less compressed pods. The ovate elliptrical basal leaves have wedge-shaped base and slender stalks. Mustard Family.

26

26. Spider Plant, *Saxifraga flagellaris,* among mosses in #22, 07-17-85. In the Canadian High-Arctic and northern Greenland, the Spider Plant usually bears a single flower and the whole plant is reddish-purple except the yellow petals (subsp. *platysepala*). The vegetative repeoduction and dispersal is by its long runners (5 each in #26) terminating in small rooting rosettes. See also # 249.

27

27. Scurvy-grass, *Cochlearia officinalis,* and Spider Plants in #22, 07-17-85. The roundish basal leaves of the Scurvy-grass have kidney-shaped or truncate base and slender stalks. The pods are ellipsoid or nearly spherical. See also # 61, 182.

28

29

28-29. Sorted polygons and circles, near Resolute Bay, Cornwallis Island, 07-17-85. In rather flat exposed areas repeated freezing and thawing results in coarser stones surrounding finer materials inside, forming sorted nets, polygons, and circles. Only occasional Tufted Saxifrage (#30), Arctic Parrya (#32), Arctic Willow, Arctic Poppy (#95), and Purple Saxifrge (#253-255) grow here.

30

30. Tufted Saxifrage, *Saxifraga caespitosa,* near Resolute Bay, 07-17-85. The flowers of the Tufted Saxifrage in the Canadian High Arctic are mostly solitary and the sepals are dark purple (subsp. *uniflora*). The broadly wedge-shaped basal leaves are three-lobed. See also # 250.

31

31. In the nearly barren rock field, the areas around bird stones and lemming dens are rather densely vegetated due to manuring effects. Here in the picture grasses, Tufted Saxifrage, Purple Saxifrage, Nodding Saxifrage and Arctic Poppy grow. The manured bird stone is covered with bright orange lichens. Near Resolute Bay, 07-14-85.

32

33

32-33. Arctic Parrya, *Parrya arctica*, in a rock field near Resolute Bay, 07-18-85. The Arctic Parrya grows in the limestone region of the Canadian Western Arctic, although it also occurs on gravel (limestone) ridges on Southampton Island, Hudson Bay. White-flowered plants are not rare. The dark puple, plump pods are often inconspicuously once constricted, somewhat curved, and with nerved valves. The somewhat fleshy leaves are all basal, narrowly oblanceolate to spatulate, entire-margined, and glabrous. The broadened, membranous, hyaline bases of the leaf-stalks are persistent, covering the basal part of the stem. Mustard Family.

34

34. In the High Arctic, well-vegetated areas are found in wet places around ponds or small lakes (#34, 41, 65, 123), along streamlets (#124, 148, 149), and on late snow-melt or seepage slopes (#34). Here, near Resolute Bay (07-14-85), in the snow-melt wet meadow around a pond on a nearly barren dry upland, are found sedges (#35), buttercups (#36), cotton-grass (#37), grasses, Yellow Marsh Saxifrage (#38), Edward's Eutrema (#39), Nodding Bladder Campion (#40), Spider Plant, and Tufted Saxifrage.

35

35. Water Sedge, *Carex aquatilis* var. *stans* (*C. stans*), Lake Hazen, 08-05-85. This sedge may be distinguished from the southern Bigelow's Sedge (#185) by its lowest bract which is longer than the inflorescence.

36

36. Sulphur-colored Buttercup, *Ranunculus sulphureus*, Resolute Bay, 07-13-85. The shallowly lobed basal leaves are rounded or short-pointed at the base. See also #15. Sabine's Buttercup, *R. sabinei*, has similar leaves but smaller flowers with gray-hairy sepals.

37. Cotton-grass, *Eriophorum triste* (*E. angustifolium* subsp. *triste*), Grise Fiord, Ellesmere Island, 06-25-83. This cotton-grass has 3 or more head-like clusters of flowers at the tip of the stem subtended by a blackish spathe. Each flower is in the axil of a blackish scale and has 3 stamens, a pistil, and many bristles which greatly elongate in fruit and become white cottony. The cottony heads have a short scabrous stalk and droop at the apex. See also #188.

38-38A. Yellow Marsh Saxifrage, *Saxifraga hirculus*. The flowering stems bear linear, entire-margined, sessile leaves. The basal leaves are also entire, narrowly spatulate or ligulate, and long-stalked. The flowers are mostly solitary. The margins of stem leaves and the stem below the flower have rusty crinkled hairs. 38: By a pond in #34, 07-18-85. 38A: on a seepage slope, Black Top Ridge, Eureka, 07-26-87.

39

40

39. Edward's Eutrema, *Eutrema edwardsii,* in #41, 08-13-85. The plants have sessile stem leaves and long-stalked basal leaves with ovate-elliptical blade. The linear pointed pods have valves with a conspicuous mid-nerve. Mustard Family.

40. Nodding Bladder Campion, *Melandrium apetalum* subsp. *arcticum* (*Silene uralensis*), in #34, 07-18-85. The plants grow in scattered tufts. The solitary flowers are nodding first but become erect later on. The purple petals are included in or slightly longer than the strongly inflated calyx with dark purple veins. The small, flat, more or less reniform seeds have a broad membranous wing. The leaves and stem are sparsely and minutely hairy. Pink Family.

41

41. Hummocky moss carpet at the margin of Small Lake (Three Mile Lake), near Resolute Bay, 08-13-85. The plants found in #34 and *Pleuropogon sabinei* (#66) also grow here. The moss is *Brachythecium albicans*.

42

42. Grise Fiord hamlet area (76°25′N), Ellesmere Island, 07-22-85. The temperatures in the hamlet ranged from 2° to 14°C (36° to 57°F) from July 19 to July 24, 1985. This area is near the northern boundary of the region which is frequently affected by the storm activity originating in the Davis Strait and Baffin Bay. Most of the storms stagnate and die in this region.

43

43. On the upper rock slopes of the mountain were Moss Campion (#44), Fragile Fern (#45), Arctic Avens, Arctic Heather in sheltered places, Arctic Poppy, Arctic Willow, Purple Saxifrage, and others. Grise Fiord, 07-24-85.

44

44. Moss Campion, *Silene acaulis*, male plant, Grise Fiord 07-23-85. Some arctic and alpine plants often grow as hemispherical cushions. The pink or purplish petals of this plant are notched at the apex. The crowded, small, flat leaves are lanceolate. Female plants have flowers with 3 long styles.

45

46

45-46. Brittle Fern or Fragile Fern, *Cystopteris fragilis*, grows in rock crevices. A few to several form a small tuft. The indusium is hood-shaped. 45: Grise Fiord, 07-24-85. 46: Near Mt. Edith Cavell, Alberta, 08-15-74. Woodsiaceae, Cliff Fern Family.

47

47. Smooth Cliff Fern, *Woodsia glabella*, Repulse Bay, 07-24-74. This small, densely tufted fern also grows usually in rock crevices, but may be found on the ground. The short leaf stalks are green and jointed near the base. The leaf segments are more or less fan-shaped. The leaf blades and stalks above the joints are not scaly nor hairy. The minute indusium is attached by its base under the sporangia and is fringed with several hairs. The plants found in rock crevices at Grise Fiord were about as small as those found on the ground at Tanquary Fiord (#138). Woodsiaceae, Cliff Fern Family.

48

49

48. Snow or Arctic Wood-rush, *Luzula nivalis* (*L. arctica*), Grise Fiord, 07-22-85. This densely tufted Wood-rush usually grows in moister places than the following species. The basal leaves are flat, about 3-4 mm wide, tapering, and often purplish at the base. The clusters of flowers are typically solitary. The bractlets of flowers are hairless. Rush Family, Juncaceae.

49. Northern Wood-rush, *Luzula confusa*, Grise Fiord, 07-19-85. This loosely or densely tufted Wood-rush often with short rhizomes occurs in dry rocky places. This and the Arctic Poppy were among the very few plants found on the summit of the mountain in #43. The narrowly channeled leaves are rolled in toward the apex. Usually 3 clusters of flowers are at the branched end of the flowering stems. The bractlets of flowers are hairy on the margins.

50

50-51. *Draba corymbosa* (*D. bellii*, *D. macrocarpa*). The rosettes of leaves are densely crowded, and the pale yellow petals are much longer than the sepals. The whole plant, including the pods, is densely hairy. 50: Grise Fiord 07-19-85. 51: Resolute Bay, 07-18-85. Mustard Family.

51

52

53

52-53. Lichens on a boulder, *Parmelia (Arctoparmelia) separata*, Grise Fiord, 07-22-85.

54

54. A dwarf shrub-moss community on a middle slope of a mountain, Grise Fiord, 07-24-85. The slope is covered with Arctic Heather mixed with Arctic Avens, Arctic Willow, Arctic Poppy, Purple Saxifrage, and the woolly moss *Rhacomitrium lanuginosum*.

54A

54A. Brown Finger Lichen, *Dactylina arctica,* among the woolly moss *Rhacomitrium lanuginosum* in a hollow by the Arctic Heather, Grise Fiord, 07-24-85. See also #273.

55

55. Dry, stony lowland near the seashore, Grise Fiord, 07-19-85. Nard Sedge, Arctic Avens, and Purple Saxifrage grow here together with some Woolly Lousewrt, Arctic Willow, and Moss Campion.

56

56. Nard Sedge, *Carex nardina*, in # 55, 07-19-85. This is a very densely tufted sedge with filiform involute leaves. The ovoid spikes are often among the leaves, composed of male flowers at the top and female flowers below with two stigmas. The pistillate scales are obtuse to rounded, brown to chestnut brown with paler margins and midveins. The perigynia are gradually narrowed to a short, shallowly bident beak.

57

57. Dry, stony lowland near the seashore, Grise Fiord, 06-27-83. Woolly Lousewort and Purple Saxifrage are conspicuous in the early season at this site, while Nard Sedge is more prominent in the later season (#55).

58

58. Woolly Lousewort, *Pedicularis lanata*, Grise Fiord 06-25-83. These plants have densely wolly inflorescence with many, large, crowded, reddish pink flowers. The lower lip of the corolla is 3-lobed. The upper lip is slightly arched, darker red at the apex, without minute lateral teeth near the apex. The plant has a deep-yellow tap-root (#78A). Arctic-alpine species, extending south to the Canadian rockies. Figwort Family, Scrophulariaceae.

59

59. Slider Fiord shore, Eureka, 07-27-87. Above the narrow, barren, sandy or gravelly beach, scattered Mountain Sorrel, Scurvy-grass. Arctic Bladder Pod, grasses, Bright Cinquefoil, Thrift (# 219), and others grow. They were all very small here.

60

60. Mountain Sorrel, *Oxyria digyna,* Eureka, 07-31-85. Frequent in moving sand, and in late snow-patch communities with mosses in the south.

61

61. Scurvy-grass, *Cochlearia officinalis,* Eureka, 07-31-85. The plant in the picture was only 2 cm (less than one inch) high. See also #182. Mustard Family.

62

62. Arctic Bladder Pod, *Lesquerella arctica,* Eureka, This silvery plant is densely covered with minute, whitish, stellate hairs. The pods are pear-shaped or almost spherical and glabrous, Mustard Family.

63

63. Grass, *Festuca baffinensis,* Eureka, 07-31-85.

64

64. Bright Cinquefoil, *Potentilla pulchella,* Slider Fiord shore, Eureka, 07-27-87. The dense basal leaves are pinnately divided into 5 leaflets with the lowest the smallest, and densely silky hairy especially beneath, the long hairs on the whitish underside extending beyond the tips of the lobes as a small whitish tuft. The small flowers scarcely project above the leaves. The flowering stems elongate later and recline. The base of the stem is densely covered with persistent, brown, broad, papery stipules. Rose Family.

65

65. A pond with Cotton-grass, *Eriophorum scheuchzeri*, on the margin and an aquatic grass, *Pleuropogon sabinei*, in the shallow water, Eureka, 07-28-85.

66

66. *Pleuropogon sabinei* (Semaphore Grass) in a pond, Eureka, 07-28-85. The grass has long, floating leaves on sterile shoots. The one-sided drooping spikelets were swaying in strong winds when the photograph was taken.

67

67. A pond with Cotton-grass, *Eriophorum scheuchzeri*, Mare's-tail (# 213), and *Pleuropogon sabinei*, Eureka, 08-02-85.

68

68. Cracked, clayey mounds with Arctic Willow in the troughs, Eureka, 07-24-87.

69

69. Cutleaf Fleabane, *Erigeron compositus,* on dry clayey ground, Eureka, 07-26-87 . The leaves are all basal and lobed into linear segments.

70

70. Red-stemmed Cinquefoil, *Potentilla rubricaulis,* Eureka, 07-26-87. This Cinquefoil differs from *P. pulchella* in its larger flowers projecting above the leaves on erect or ascending flowering stems which are often tinged with red. It grows mainly in open clayey or gravelly places, while the Bright Cinquefoil grows chiefly in sandy places near the sea.

71

71. Large, high-centered, ice-wedge polygons, Eureka, 08-01-85. The clay hummocks are covered with Arctic Avens (*Dryas integrigolia*) and *Kobresia myosuroides*. Other plants also growing here include Arctic Poppy, Arctic Willow, Mountain Sorrel, Viviparous Knotweed (#221), and Hairy Lousewort (# 77). Occasionally Arctic Heather occurs in the channels between the hummocks and sometimes covers the hummocks in depressions.

72

72. *Dryas integrifolia-Kobresia myosuroides* community on a high-centered ice-wedge polygon, Eureka, 08-01-85. *Kobresia myosuroides* (*K. bellardii*) is a densely tufted sedge-like plant with leaves very much like those of Nard Sedge (#56) but shorther than stems with slender inflorescence. The spikelets are 2-flowered, the upper staminate and the lower pistillate. The thin, shiny scale (perigynium) which closely encloses the female flower and fruit is open along the margin.

73

73. Arctic Willow hummocks along a wet channel between two gentle slopes of ice-wedge polygons, Eureka, 07-28-85. The plants growing in the wet channel include Arctic Lousewort (# 76), Viviparous Knotweed (#221), *Eutrema edwardsii* (#39), *Ranunculus sulphureus* (#36), Alpine Cress (#25), and Arctic Poppy.

74

75

74. *Melandrium triflorum*, on dry clayey ground near the seashore, Eureka, See #18.

76

75-76. Arctic Lousewort, *Pedicularis arctica* (*P. langsdorfii* subsp. *arctica*). 75: among *Cassiope mertensiana*, near Mt. Edith Cavell, Alberta, 08-14-74; 76: between Arctic Willow hummocks, Eureka, 07-24-87. The large flowers on leafy stems have a prominently arched upper lip with two small acute teeth below the truncate apex. The style is protruding. The calyces, bracts, the upper part of the stem, and the bases of the upper leaves are more or less woolly hairy. The bases of the basal leaf stalks are expanded and persistent.

77

77-78. Hairy Lousewort, *Pedicularis hirsuta,* Grise Fiord, 07-19-85. The rather short upper lip of the pale pink corolla is straight and lacks teetn near the tip. The style is included in the upper lip. The plant is woolly hairy and has a whitish tap-root. The bases of the basal leaf stalks are expanded and persistent.

78A. Woolly Lousewort, *Pedicularis lanata,* Grise Fiord, 06-24-83. The plant has a deep-yellow tap-root. See also 58.

78

78 A

79

79. On the moist stony streamsides, Dwarf Fireweed grows with some Purple Saxifrage and Arctic Willow. Eureka, 07-24-87. The mountain top of Black Top Ridge, 800 m (2600ft) in elevation, is a level stony barren with only trace of Arctic Poppy and Brook Saxifgage. The steep rock slopes support only occasional Saxifrages, Arctic Poppy, Mountain Sorrel, Arctic Willow, and *Potentilla vahliana* (#91). More plants grow on the moister upper level areas.

80

80. Dwarf Fireweed, Broad-leaved Willow-herb or River Beauty, *Epilobium latifolium* (*Chamaenerion latifolium*), Eureka, 07-24-87. The flowers have an inferior ovary which becomes a many-seeded capsule. The seeds have a tuft of hairs at one end. Arctic and alpine species, extending south to Colorado (*Chamerion subdentatum*).

81

81. Arctic Hare is common is Eureka and other places. Bathurst Island, 06-29-87. Photo by Bezal Jesudason.

82

83

82-83. Arctic Heather, *Cassiope tetragona,* Eureka, 07-24-87. The Arctic Heather grows in hummocks where evidently the drift snow accumulates in winter above the lower slopes with Arctic Avens, Nard Sedge, Arctic Willow, and Saxifrages.

84

85

84-85. Although rare in the High Arctic, the Alpine Bilberry, *Vaccinium uliginosum* var. *alpinum*, occasionally grows in Arctic Heather hummocks apparently benefiting from the snow cover there in winter. No fruits were found on these plants. Tanquary Fiord, 08-11-87. Tanquary Camp is the northernmost known Canadian locality for this plant.

86

87

36-87. Pond Inlet (72°43′N), northern Baffin Island, 07-05-83. In contrast to most high-arctic areas, the low rolling, middle-arctic land around Pond Inlet is covered with continuous vegetation, including heathlands dominated by Arctic Heather, *Cassiope tetragona*.

88

88. Pond Inlet, middle-arctic Beffin Island, 07-05-83. Here in a drier place, Lichens fill the gaps between the flowering plants, forming continuous vegetation, which include yellow Snow Lichen (#266), black thread-like Lichen (#271), Woolly Lousewort, (#58), Arctic Heather (#200), Alpine Bilberry (#203), Arctic Avens (#233), Arctic Willow (# 240), Purple Saxifrage (#253), Northern Labrador-tea (#5), and Lapland Rosebay (#5). Such dense growth of lichens was not found in the High Arctic where precipitation is very low.

89

90

89-90. Dry, gravelly, middle slope of Black Top Ridge. Eureka with widely scattered Alpine Arnica, *Arnica alpina* subsp. *angustifolia*, 07-28-85, 07-24-87. See also #166.

91

92

91-92. Vahl's Cinquefoil, *Potentilla vahliana*, cn a dry rocky slope, Black Top Ridge, Eureka, 07-28-85, 07-29-87. These densely tufted plants with 3-parted leaves have branches covered with old brown stipules and the remnants of leaf stalks with long spreading hairs. The long haire on the whitish or yellowish undersides of the leaflets form small tufts at the tips of the teeth. The flowers are solitary.

93

93. Alpine Avens, *Geum rossii (Acomastylis rossii)*, Black Top Ridge, Eureka, 07-24-87. Although rare in the High Arctic, this species is rather common in Alaska and the American Rockies (see #237). These plants grow in dense tufts or mats. The long basal leaves have more or less lobed or toothed segments with hairy margins. The flowers are mostly solitary. The sepals, bractlets, and top-shaped calyx-tubes are short-hairy as are the stems with few greatly reduced leaves. The ovaries are densely hairy, but the styles are hairless.

94

95

94-95. Arctic Poppy, *Papaver radicatum,* on an upper rock slope, Black Top Ridge, Eureka, 07-26-87. Arctic Poppy grows here with Saxifrages and others.

96

96. A high-arctic red moss, *Bryum cryophilum,* along a streamlet on Black Top Ridge, Eureka, 07-29-87. It is speculated that some high arctic plants are always red as an adaptation to the high ultraviolet radiation.

97

98

97-98. Arctic Fleabane, *Erigeron eriocephalus (E. uniflorus* subsp. *eriocephalus*), on a moist rock slope by a streamlet, Black Top Ridge, Eureka, 07-24-87. The purple involucral bracts of the flower heads are spreading and curving upwards, and densely covered with soft, pale hairs. The stems have spreading, soft hairs, and the leaves are also soft-hairy. Widely distributed in the Arctic.

99

99. Northern or Birdfoot Buttercup, *Ranunculus pedatifidus* var. *leiocarpus* (*R. affinis*), Black Top Ridge level area, Eureka, 07-26-87. The basal leaves are long-stalked, cordate at the base, rounded in outline, palmately cleft into narrow lobes. The stem leaves are sessile or nearly so, divided into linear segments. The greenish-yellow sepals, flowering stems, and leaf margins and stalks are with soft, whitish hairs. The petals are 8-10 mm long.

100

100. Northern or Birdfoot Buttercup, *Ranunculus pedatifidus* var. *leiocarpus*, Black Top Ridge level area, Eureka, 07-26-87 .

101

101. *Taraxacum phymatocarpum*. Black Top Ridge level area, Eureka, 07-26-87. This dwarf, arctic dandelion has light-yellow flowers. The leaves are only slightly lobed, toothed, or nearly entire.

102

102. A fell-field with Prickly Saxifrage, *Saxifraga tricuspidata*, Black Top Ridge level area, Eureka, 07-29-87. Other plants growing on this level area (elevation about 500-550 m, 1640-1800 ft) include those illustraed in #25, 26, 27, 30, 36, 40, 48, 91, 95, 99, and small patches of Arctic Willow.

103

103. Black Top Ridge talus slope. On the level area above this talus slope, at elevation about 600m (2000ft), Alpine Chickweed, Arctic Poppy, Nodding Saxifrage (#251), Purple Saxifrage (#253), Spider Plant (#26), and Scurvy-grass (#27) grow. On a nearby ridge top at 600 m elevation, surrounded by barren gravelly plateau, a seepage streak was found where many Tufted Saxifrages were in bloom together with some Arctic Poppies and a few Spider Plants, Alpine Saxifrage, and Alpine Foxtail. No Arctic Willow or other woody species was found.

104

104. Alpine Chickweed, *Cerastium alpinum,* Omingmak Mountain, near Lake Hazen, 08-03-87. The plant is densely covered with long white hairs. The flowers have notched petals and 5 styles. The leaves are oblanceolate to oblong.

105

105. Mosses predominate in the late snowbed area below the mountain top, Black Top Ridge, Eureka, 07-29-87.

106

107

106-107. Ice-caps, snowfields, and glaciers in Princess Margaret Range, southeastern Axel Heiberg Island, near Eureka, Ellesmere Island. 07-23-87.

108

108. Ice-caps and glaciers in the mountains between Tanquary Fiord and Lake Hazen, Ellesmere Island, 07-30-87.

109

110

111

109-111. Macdonald River delta, Tanquary Fiord (81°25′N), 08-10-85. The northernmost part of Ellesmere Island, north of Greely and Archer fiords, is known as Grant Land. This is a mountainous country with many ice-caps and glaciers, and is generally an area of folded sedimentary rock and calcareous soil. Osborn Mountains rise abruptly to sharp mountain ridges 1400 to 1500 m (4500-5000ft) high on the northwest side of the Tanquary Fiord over the extensive river bed (#110). Although Arctic Avens, Arctic Willow, Capitate Lousewort (#256), and Mountain Sorrel were growing on the steep slopes at the edges of the flat gravel terraces (#111), only very occasional *Melandrium triflorum* (#18), Arctic Bladder Pod (#62), and *Potentilla rubricaulis* (#236) were found on the terraces. The summers of the Tanquary Fiord-Lake Hazen area are warm for the high latitude. The July mean temperature at Tanquary for the period of 1963-1965 and 1967 was 5.8°C (42.5°F). From August 6 to August 13, 1987 temperatures at Tanquary Camp ranged from 5.5°C (42°F) to 12.2°C (54°F).

112

112. Tanquary Fiord shore near the Macdonald River, 08-10-85. Salt-marsh plants (#116, 117) grow here with associated seashore plants (#61, 175, 186, 219).

113

113. A lagoon bordered by Bear Sedge,and with Floating Buttercup (# 118) and *Dupontia fisheri* (#119) in the water, Tanquary Fiord, 08-10-87.

114

114. Densely tufted Bear Sedge, *Carex ursina,* Tanquary Fiord shore, 08-10-87. Tanquary Camp is the northernmost locality for this sedge on Ellesmere Island.

115

115. A salt marsh along the Tanquary Fiord shore on the Macdonald River delta, 08-11-85. Creeping Alkaligrass and Bear Sedge are well developed here.

116

116. Creeping Alkaligrass, *Puccinellia phryganodes*, and scattered, densely tufted Bear Sedge in the salt marsh (# 115).

117

117. Bear Sedge, *Carex ursina*, Tanquary Fiord, 08-10-87. These plants do not have long, creeping rhizomes like the Curved Sedge (# 186). The narrow leaves are channeled or involute. The head is among the leaves and consists of a single spike with male flowers at the base and female flowers with two stigmas above. The scales are ovate, obtuse, brown to dark purplish brown, lighter on the midvein and on the narrow margins. The perigynium is yellowish brown, rounded and abruptly minutely beaked at the tip, short-stalked at the base.

118

118. Floating Buttercup, *Ranunculus hyperboreus,* growing in aquatic moss in a lagoon, Tanquary Fiord, 08-10-87. The long-stalked small leaves are deeply 3-lobed. See also #143.

119

119. Grasses, *Dupontia fisheri* and Alpine Foxtail (*Alopecurus alpinus*) in a wet place near the seashore, Eureka, 08-01-85. The narrow and contracted panicles of *Dupontia fisheri* have shiny, bronze colored, 2- to 4-flowered spikelets. The 3-nerved glumes are shorter than the spikelet. and the lemmas are appressed hirsute along the keel and margins. Tanquary Camp is the northernmost Canadian locality for this stoloniferous marsh grass on arctic seashores.

120

120. A mountain slope with Arctic Avens and some Arctic Willow on hummocks, at about 600m (2000ft) elevation, just above the Omega Lakes, Tanquary Fiord, 08-11-87. Occasional *Chrysanthemum integrifolium* (#155) is found on this slope.

121

121. Omega Lake near Redrock Glacier from Viking Ice Cap, Tanquary Fiord, 08-11-87. The lake, at an altitude of 340m (1115ft), is about 10km (6 miles) northeast of Tanquary Camp.

122

122. Viking Ice Cap and glaciers, Tanquary Fiord, 08-10-85.

123

123. Omega Lakes and Mt. Bowman with an ice field across the Macdonald River, Tanquary Fiord, 08-11-87. There are well-developed sedge meadows at the lakes.

124

124. Omega Lake outlet bordered by moss hummocks, Tanquary Fiord, 08-11-87. *Cerastium regelii* (#125), *Equisetum variegatum* (#146), *Arctagrostis latifolia* (#150), Alpine Foxtail (#220), Nodding Saxifrage (#251), Nodding Bladder Campion (#40), Yellow Marsh Saxifrage (#38A), and *Cardamine pratensis* grow here.

125

125. Regel's or Polar Chickweed, *Cerastium regelii*, by the Omega Lake outlet, 08-11-87. These plants form dense mats or convex cushions. The closely placed leaves are short and broad, somewhat fleshy, and hairless. Long-internoded, terminal shoots may be induced under warm, humid conditions. Species mainly north of 70°N.

126

126. Kettle Lake and Mt. Timmia (875m, 2870ft) beyond the Macdonald River, Tanquary Fiord 08-09-85.

127

127. Arctic Avens, *Dryas integrifolia*, hummocks on the lower slope of Mt. Timmia, 08-10-85. The Arctic Avens and Arctic Willow are abundant in the Tanquary area.

128

128. Arctic Avens hummocks with Arctic Willow, Capitate Lousewort (# 256), and Viviparous Knotweed (#221) on Kettle Lake basin slope, 08-07-87.

129

130

129. A fell-field with Prickly Saxifrage, *Saxifraga tricuspidata*, near the summit of Mt. Timmia, 08-09-85. The higher up one goes woody species such as Arctic Avens, Arctic Willow, and Arctic Heather become less frequent and herbaceous species become more and more common, including Arctic Poppy, Saxifrages (#250-255), and Wood-rushes (#48-49). Near the summit the woody species drop out, and at the summit only few herbaceous species grow, including Arctic Poppy, Nodding Saxifrage (#251), and Purple Saxifrage (#253).

130. Alpine Saxifrage, *Saxifraga nivalis*, near the summit of Mt. Timmia, 08-10-85. These plant were about 10 cm (4 inches) tall. The stems are hairy and all the flowers are short-stalked. The white, short petals are about as long as the sepals. The somewhat fleshy, basal leaves are roundish and coarsely toothed.

131

131. Arctic Willow, *Salix arctica*, Tanquary Fiord, 08-07-87.

The author near the Omega Lakes, Tanquary Fiord, 1987.

132

132. *Saxifraga tenuis* (*S. nivalis* var. *tenuis*) is smaller than *S. nivalis*. The plants in the picture were only about 5 cm (2 inches) tall. The lower flowers have elongated stalks. The short pink petals are slightly longer than the sepals. Found in a wet gravelly place by a brook, Eureka, 07-28-85.

133

133. Green Valley, Tanquary Fiord, 08-07-87. There are several ponds in this valley. In the sedge meadows around them are Spider Plant, Yellow Marsh Saxifrage, *Juncus biglumis, Polygonum viviparum, Ranunculus sulphureus, Melandrium apetalum, and others. Ranunculus hyperboreus* grows in the water.

134

134. The May Creek area, Tanquary Fiord, 08-07-87. The slope in the foreground is covered with Arctic Heather hummocks and Arctic Avens-Arctic Willow hummocks.

135

135. Hummocks on a slope along May Creek, 08-12-87. Arctic Heather (*Cassiope tetragona*) grows on the sides of the hummocks; Arctic Avens (*Dryas integrifolia*) and Arctic Willow grow on the tops. Apparently the sides of the hummocks are more protected by snow.

136

136. Arctic Avens-Arctic Willow community on an upper slope, at about 600m (2000ft) elevation, along May Creek 08-07-87.

137

137. Arctic Avens-Arctic Willow community on the lower slopes along May Creek, 08-07-87.

138

139

138-139. *Woodsia glabella* on the ground in an area with Arctic Avens, Purple Saxifrage, and others at about 550m (1800ft) elevation, along May Creek, 08-12-87. The coin is 18mm across. Woodsiaceae, Cliff Fern Family.

140

140. Musk-oxen are frequently seen in the High Arctic. Polar Bear Pass near Bracebridge Inlet, Bathurst Island, 06-29-87. Photo by Bezal Jesudason.

141. A tarn and Omingmak (Musk-ox) Mountain, near Lake Hazen, Ellesmere Island, 08-05-85. Temperatures at Hazen Camp (81°49′N) ranged from 2.2°C (36°F) to 8.3°C (47°F) in early August 1985 and it snowed on the mountains on August 6, 1985 and August 1, 1987. The summers are warm for the high latitide. The July mean temperature was about 6.7°C (44°F) in the 1961-1963 period. However, it is very cold in winter. In 1957/58 winter the lowest temperature recorded was -55.8°C (-68.5°F). Here at Tarn 10 (Duck Lake), Cotton-grass, Water Sedge, Alpine Foxtail and Two-flowered Rush (#24) grow on the shore; and Mare's-tail, Floating Buttercup, and *Pleuropogon sabinei* (#66) grow in the shallow water.

141

142

142. Mare's-tail, *Hippuris vulgaris,* and Floating Buttercup, *Ranunculus hyperboreus* in #141, 08-06-87. See #213 for Mare's-tail.

143

143. Floating or Arctic Buttercup, *Ranunculus hyperboreus,* in #141, 08-02-87. The leaves are deeply 3-lobed, with the lobes rounded at the ends and the lateral lobes often shallowly notched. The small flowers have 3 petals and 3 sepals.

144

144. Floating Buttercup growing in moss under water, Skeleton Lake (Loon Lake) edge, near Lake Hazen, 07-31-87. See also #229.

145

145. Omingmak Mountain and a tarn, near Lake Hazen, 08-05-85. The Tarn 12 is bordered by a broad sedge meadow of Water Sedge and Cotton-grass.

146

146. Hazen Camp Pond surrounded by Water Sedge (#35) and Cotton-grass.

147

147. Horsetail, *Equisetum variegatum,* and Arctic Avens by the streamlet in #148, 08-05-85. The sheath teeth of the Horsetail are dark brown to black and scarious-margined. Cone bearing plants were not seen.

148

148-149. A sedge meadow along a streamlet from the Tarn 12 and draining the southeast slope of Mt. McGill, near Lake Hazen, 08-05-85.

150. Grass, *Arctagrostis latifolia*, among Water Sedge, *Carex aquatilis* var. *stans*, along the streamlet in #149, 08-05-85.

149

150

151

152

151. Arctic Avens hummock slope, southeast of Tarn 10, 08-05-85. Arctic Avens, *Dryas integrifolia,* forms hummocks over wide areas. Arctic Heather often covers the sides of these hummocks.

152. Arctic Avens-Nard Sedge community with Arctic Bladder Pod (#62) and *Braya humilis* subsp. *arctica* (# 156), Lake Hazen, 08-30-87.

153

153. Arctic Avens, *Dryas integrifolia,* on clay hummocks, Lake Hazen, 08-85. Among other plants found here are Arctic Willow, Purple Saxifrage, Arctic Poppy, Arctic Bladder Pod, Mountain Sorrel, and *Draba corymbosa* (#50, 51).

154

155

154. *Braya purpurascens* and *Kobresia myosuroides* on Arctic Avens hummocks, Lake Hazen, 08-04-85. *Braya purpurascens* (*B. glabella* var. *purpurascens*) has dark purple, erect or arching stems with stellate hairs. The plump, elliptic-oblong pods are 8-10mm long, glabrous or slightly hairy. *Kobresia myosuroides* is a densely tuffted, sedge-like plant with filiform leaves and short, linear, spike-like inflorescence. See #72.

155. Entire-leaved Chrysanthemum, *Chrysanthemum integrifolium,* in Arctic Avens hummocks, Lake Hazen, 08-05-85. The mostly basal leaves are tufted, linear, entire, and hairy on the margins. The stems are woolly above. The involucral bracts have broad, dark brown margins toothed towards the tips. Its main range of distribution is south of the 75°N latitude.

156

156. *Braya humilis* subsp. *arctica,* Lake Hazen, 07-31-87. The plant is branched at the base and hairy. The spreading, linear pods are about 1 mm wide, slightly swollen at intervals, and minutely hairy. The style has a distinctly 2-lobed stigma. The basal leaves are oblanceolate, entire or toothed.

157

157. Mt. McGill (altitude 1070m, 3500ft), near Lake Hazen, 08-03-87. The vegetation on the moist north slope changes from the lower slope dominated by Arctic Avens and Arctic Willow to the middle slope where the Arctic Heather predominates, then to the upper slope without woody speceis. The herbaceous species here include species of *Saxifraga* (*S. caespitosa*, *S. cernua*, *S. flagellaris*, *S. nivalis*, *S. oppositifolia*, *S. rivularis*, *S. tenuis*, *S. tricuspidata*), *Luzula nivalis*, *Oxyria digyna*, *Papaver radicatum*, *Potentilla hyparctica*, *Ranunculus sulphureus*, and *Cerastium arcticum*. Near the summit, Arctic Poppy and Saxifrages grow in areas between the boulders.

158

158. The middle north slope of Mt. McGill with Arctic Heather, *Cassiope tetragona*, 08-01-87.

159

159. Brook Saxifrage, *Saxifraga rivularis*, on the upper north slope of Mt. McGill, 08-01-87. Here about 10 plants were found in a dense cluster in moss among rocks. They were 1.5-3cm tall. The basal and stem leaves are flabelliform with wedge-shaped or subcordate base and deeply 3-to 5-lobed. The bases of the basal leaf stalks are expanded to broad marginal wings. The herbage is often red colored. The flowering stems are densely hairy above with long, soft, more or less tangled hairs, and bear 1-3 small flowers on similarly haired stalks. The petals are white or pale pink. Circumpolar, arctic, subarctic, and alpine species.

160

161

160-161. Nonsorted polygons and Dryas hummocks on the Arctic Coastal Plain of Banks Island near Sachs Harbour. It appears that such polygons may be initiated by desiccation and hummocks are formed primarily when the soil is domed up through frost action. Around Hazan Camp and Eureka, Dryas hummocks are well developed in ice-wedge polygons (#71).

162

163

Common Arctic and Subarctic Plants

Common arctic plants not shown in the previous pages are included here. Seventy species of flowering plants are arranged generally in the alphabetical order of their family names. They are preceded by ferns and club-mosses, and followed by lichens.

The Ferns and Club-mosses
See also #45-47, 138

164

163-164. Fragrant Wood Fern, *Dryopteris fragrans*, Anaktuvuk Pass, Alaska. This is a widely distributed arctic and alpine fern in rock screes and cliff crevices. The new fronds in dense crowns are accompanied by curled withered fronds. The short leafstalks are densely scaly. The indusium is orbicular-reniform. Dryopteridaceae, Wood Fern Family.

165

165. Fir Club-moss or Mountain Club-moss, *Lycopodium selago*, Anaktuvuk Pass, Alaska. The plants are dichotomously branched and often bear bulbils in the upper leaf axils. The reniform sporangia are borne in the axils of ordinary leaves. Widely distributed arctic and alpine species. Lycopodiaceae, Club-moss Family

Asteraceae (Compositae), Sunflower Family #166-171
See also #21, 69, 97-98, 101, 155

166

166. Alpine Arnica, *Arnica alpina,* Churchill, Manitoba. For the stunted and scattered growth in the High Arctic see #89, 90. These softly hairy plants have a few pairs of opposite, sessile stem leaves and stalked basal leaves which are tapering to the tip and conspicuously nerved. The involucral bracts are densely white-hairy at the base. Found on dry, rocky and gravelly slopes.

167

167. Dwarf Hawksbeard, *Crepis nana,* Whistlers Mountain, Alberta. This species found on scree slopes and gravels has small, short-stalked flowerheads crowded among the leaves. The involucre is cylindrical. It ranges, in North America, mainly from 70°N, south to 50°N and Colorado (*Askellia nana*).

168

168. Mountain Fleabane, *Erigeron humilis* (*E. unalaschkensis*), near Mt. Edith Cavell, Alberta. The involucral bracts are dark purple, appressed, and densely covered by woolly hairs with dark purplish corsswalls. A low-arctic, subarctic, and alpine species south to Colorado.

169

169. Arctic Daisy, *Chrisanthemum arcticum*, Churchill, Manitoba. The solitary heads have white ray flowers and dark brown involucral bracts. The fleshy, wedge-shaped basal leaves are wavy-toothed or cleft at the apex. Subarctic coastal plant from Alaska to Hudson and James Bays in North America.

170

170. Seashore Chamomile, *Matricaria ambigua*, Holman, Victoria Island. These seashore plants of the Low Arctic have finely divided leaves and involucral bracts with dark brown, scarious, toothed margins.

171

171. Marsh Ragwort, Marsh-fleabane or Mastodon Flower, *Senecio congestus*, Cambridge Bay, Victoria Island. These hairy plants found on moist seashores and at the edges of ponds have hollow stems densely covered by brownish yellow hairs. Several to many flower-heads are in congested clusters. Arctic and subarctic in distribution.

172

Boraginaceae, Borage Family

172. Sea Lungwort or Oysterleaf, *Mertensia maritima,* Churchill, Manitoba. This spreading seashore plant with fleshy glaucous leaves bears small purplish-pink or bluish bell-shaped flowers and spongy fruits. Wide-ranging, mainly low-arctic and subarctic.

173

Caryophyllaceae, Pink Family #173-181
See also #18, 40, 44, 104, 125

The genera usually included in the family but with separate sepals and clawless petals are now segregated as Alsinaceae, Chickweed Family, which includes #104, 125, 174-181.

173. A sandy beach with mats of Seabeach Sandwort, Churchill, Manitoba.

174

174. Seabeach Sandwort or Seapurslane, *Honckenya peploides* (*Arenaria peploides*), Pangnirtung Fiord, Baffin Island. The crowded yellowish-green leaves are rather large, fleshy, and opposite. The stamens and petals are inserted on the yellowish disk. A mainly low-arctic species on sandy seashores.

175

175. Low Chickweed, *Stellaria humifusa,* Barrow, Alaska. These densely matted, hairless, seashore plants have flowers with deeply cleft petals and 3 styles. The trailing stems have opposite, sessile, ovate to elliptical, somewhat fleshy leaves which remain attached in a withered condition making the mat brownish. The northernmost locality for this species is Tanquary Camp, Ellesmere Island.

176

176. The Low Chickweed with very deeply cleft petals and 3 styles commonly grows among the Creeping Alkaligrass (*Puccinellia phryganodes*) along the arctic seashores subject to flood at high tide. Barrow Spit, Alaska.

177

177. Tufted Pearlwort, *Sagina caespitosa,* Iqaluit (Frobisher Bay), Baffin Island. These very small plants in compact hemispherical tufts grow in moist gravelly places in the Low Arctic of eastern Canada and West Greenland. The abundant, small flowers have 5 petals, 5 sepals, and (4 or) 5 styles. The petals are slightly longer than the sepals. The short flower-stalks and the bases of the sepals are minutely hairy. The leaves are slender and persist at the base of tufts when withered.

178

179

180

181

178-179. Tufted Pearlwort, *Sagina caespitosa*, Iqaluit, Baffin Island. The sepals are green and ovate with often purple blunt tips.

180-181. Reddish Sandwort, *Minuartia rubella* (*Arenaria rubella*), Resolute Bay, Cornwallis Island. The sepals are reddish or purplish, lanceolate, acuminate, distinctly 3-nerved, equalling the petals, and slightly pubescent. The flower-stalks are densely glandular hairy. The leaves are linear-subulate, acute, and strongly 3-nered. The styles are 4 or 5 in these specimens.

182

Brassicaceae (Cruciferae), Mustard Family #182-184
See also #19, 25, 27, 32, 33, 39, 51, 61, 62

182. Scurvy-grass, *Cochlearia officinalis,* Barrow, Alaska. These somewhat fleshy plants form rosettes of roundish leaves with kidney-shaped or truncated bases and slender stalks in the first year. The small white flowers with 4 petals appear among the leaves in the second year. For the ellipsoid or nearly spherical pods see #61. A coastal species throughout the Arctic.

183

183. Pallas's Wallflower, *Erysimum pallasii,* Arctic Bay, Baffin Island. The flowers with 4 purple petals and 4 purple sepals are borne on the very short flowering stem which elongates later.

184

184. Pallas's Wallflower, Holman, Victoria Island. The long linear pods are curved. The plant is densely covered with appressed, forked hairs attached in the middle. This arctic and alpine species grows in calcareous stony places, extending south to the Canadian Rockies.

185

Cyperaceae, Sedge Family #185-190
See also #35, 37, 56, 114, 117

185. Bigelow's Sedge, *Carex bigelowii,* Iqaluit (Frobisher Bay), Baffin Island. The inflorescence consists of one terminal male spike and lower 1-3 cylindrical, nearly sessile, erect female spikes with 2 stigmas. The lowest bract is usually shorter than the inflorescence. This arctic sedge found in moist or rather dry places is not known from western N. America.

186

186. Curved Sedge, *Carex maritima,* Holman, Victoria Island. The dwarf plants with often curved leaves and short, erect or curved stems grow chiefly on sandy, arctic seashores, but may also occur inland. They have long creeping rootstocks. The globose heads are composed of several, sessile, bisexual spikes densely crowded together. Each spike is staminate at the top, pistillate with 2 stigmas at the base. The perigynia are yellowish brown to dark purple at maturity, gradually narrowed to a slender beak.

187

187. Hoppner Sedge, *Carex subspathacea,* Repulse Bay, N.W.T. This is a dwarf species of arctic coastal salt marshes. It has creeping rootstocks, forming sods. The leaves are channelled toward the base and involute toward the tip. The lowest bract is dilated at the base, ensheathing the base of the lowest female spike which is ellipsoid-oblong and few-flowered with 2 stigmas. The Canadian dime shown is 18 mm across.

188

188. Tall or Narrow-leaved Cotton-grass, *Eriophorum angustifolium*, Repulse Bay, N.W.T. There are 3 or more spikes (heads) on erect, ascending, or drooping stalks at the apex of the stem. The scales of the spikes are grayish to brownish. These chiefly low-arctic and subarctic plants with creeping root-stocks grow gregariously in marshes and have flat leaves near the base. For the arctic plants see #37.

189

189. Scheuchzer's or Arctic Cotton-grass, *Eriophorum scheuchzeri*, Iqaluit, Baffin Island. There is only one globose head at the apex of the stem. The few leaves near the base are filiform. The plants grow in wet places around the arctic and subarctic ponds and lakes.

190

190. Sheathed Cotton-grass, *Eriophorum vaginatum* subsp. *spissum* (*E. spissum*), Pond Inlet, Baffin Island. This cotton-grass with solitary heads is densely tufted, forming compact, rounded tussocks. The scales at the base of the head are strongly reflexed, ovate-acuminate, lead-gray or darker with pale margins. The similar sedge (*E. vaginatum*) in western N. America has rather uniformly lead-gray scales on the head.

191

Diapensiaceae, Diapensia Family

191. Diapensia, *Diapensia lapponica*, Pangnirtung, Baffin Island. This dwarf shrub forms cushions in exposed rocky and gravelly places. The flowers are solitay on stiff stalks. The bell-shaped corolla is 5-lobed, and the 5 stamens are attached to the corolla tube. The crowded, small, evergreen, narrow leaves are spatulate. Southern arctic, subarctic, and alpine.

192

Empetraceae, Crowberry Family

192. Black Crowberry, *Empetrum nigrum*, Anaktuvuk Pass, Alaska. This is a creeping or spreading evergreen shrub with black, edible, juicy, berry-like fruits. The crowded, short-stalked, spreading, linear leaves are about 5 mm long and have a groove on the lower surface formed by the strongly rolled-under margins. The leaf margins are not minutely toothed as in the Mountain Heather (# 10). An arctic, subarctic, and alpine species.

193

193. Black Crowberry, Lake Agnes area, Alberta, 06-13-83. The flowers are small, axillary, unisexual or bisexual; stamens with dark purple anthers are much longer than the 3 spreading purplish petals and 3 greenish sepals.

194

Ericaceae, Heath Family #194-205
See also #4, 5, 8-11

194. Mats of Alpine Bearberry among gray coral-like lichens (#267) in a gravelly place, Auyuittuq National Park, Baffin Island.

195

195. Alpine Bearberry, *Arctostaphylos alpina* (*Arctous alpina*) in bloom in early spring, with persistent withered leaves and new leaves beginning to unfold,

196

196. Alpine Bearberry with the leaves turned deep red and with ripe, black, berryk-like fruits in autumn, Dempster Highway, Yukon.

197

197. Pangnirtung Pass trail with Arctic Heather and gray, coral-like lichens (#267) between the boulders, Auyuittuq National Park, Baffin Island.

198

198. Arctic Heather, *Cassiope tetragona,* by a lichen-clad boulder in #197. See also #200.

199

199. Red-fruited Bearberry, *Arctostaphylos rubra* (*A. alpina* subsp. *rubra, Arctous rubra*) in autumn, Summit Lake, British Columbia. The leaves are thinner, less deeply wrinkled, with longer stalks, and more readily deciduous than those of the Alpine Bearberry. Common in the western Subarctic.

200

200. Arctic Heather, *Cassiope tetragona,* Repulse Bay, N.W.T. This circumpolar, arctic species has evergreen scale-like leaves arranged in four distinct rows. They are deeply grooved on the back.

201

201. Labrador-tea, *Ledum groenlandicum,* Jasper National Park, Alberta. This is a subarctic, evergreen shrub with flowers in crowded terminal clusters subtended by the uppermost leaves. The corolla is very deeply 5-lobed and the lobes are spreading. The leaves are narrowly ellitic, leathery, with strongly rolled-under margins, and densely covered with rusty brown woolly hairs on the underside. Common in boggy places and in Black Spruce forests.

202

202. Alpine Azalea, *Loiseleuria procumbens,* Salmon Lake, Alska. The prostrate, much-branched, evergreen, dwarf shrub often forms mats. The flowers are in few-flowered terminal clusters, borne on short stalks from the axils of crowded upper leaves. The pink, bell-shaped corolla is deeply 5-lobed and there are 5 stamens. The calyx lobes and flower-stalks are dark red. The leathery opposite leaves are small, elliptic, rounded at the tips, densely matted-hairy on the undersides between the rolled-under margins and the broad midvein. A low-arctic, subarctic, and alpine species, extending south to the Canadian Rockies and the northern Appalachians.

203

203. Alpine Bilberry or Arctic Blueberry, *Vaccinium uliginosum* var. *alpinum* (*V. uliginosum* subsp. *microphyllum*, *V. gaultheroides*), Auyuittuq National Park, Baffin Island. This is a dwarf deciduous shrub with very small, rounded, entire-margined, net-veined leaves. The corolla lobes are usually 4 and reflexed. Arctic, subarctic, and alpine in distribution.

204

204. Mountain Cranberry, *Vaccinium vitis-idaea* var. *minus*, Churchill, Manitoba. These dwarf prostrate subshrubs have shiny, leathery, evergreen leaves which are pale and dark-spotted on the undersides, and white or more commonly pink flowers in short terminal clusters. The corolla is 4-lobed.

205

205. Mountain Cranberry with lichens, at timberline, Tungsten, N.W.T. The red, rather acid berries make good cranberry sauce by adding sugar.

206

Fabaceae (Leguminosae), Pea Family
#206-212
See also #6

206. Alpine Milk-vetch, *Astragalus alpinus*, Iqaluit (Frobisher Bay), Baffin Island. The slender, spreading and freely branching stems with pinnately compound leaves form a mat. The whitish-purplish, fragrant flowers in crowded short clusters produce densely black-hairy pods. Arctic, subarctic, and alpine in distribution.

207

207. Holman, Victoria Island.

208

208. Arctic Oxytrope, *Oxytropis arctica*, in #207. The 2-5 large, purple flowers are at the end of long, soft-hairy flowering stems. The banner (standard) and wings are longer than the keel which is sharp-pointed. The calyx with linear-lanceolate teeth is about 10 mm long, and densely covered with both black and white hairs. The small leaflets and leaf stalks are densely white-soft-hairy, and the papery white stipules, especially their margins, are long-white-hairy. This tufted arctic species occurs in western Canada and Alaska.

209

209. Silvery Oxytrope, *Oxytropis arctobia* (*O. nigrescens* subsp. *arctobia*), Cambridge Bay, Victoria Island. These densely hairy plants are very densely tufted, often forming flat or hemispherical cushions. The purple flowers are mostly solitary on very short flowering stems with white-silky hairs. The calyx with linear-lanceolate teeth is about 7 mm long, and densely covered with both black and white hairs. The pods are oblong-cylindrical and covered with both black and white hairs. The small leaflets, leaf-stalks, and the straw-colored stipules are densely clothed with white-silky hairs These plants grow in the limestone region of southern arctic Canada.

210

210. Bladder Locoweed or Inflated Oxytrope, *Oxytrois podocarpa*, Iqaluit, Baffin Island. This species is somewhat similar to the Silvery Oxytrope, but the less densely tufted plants (#211, 212) are thinly hairy; the flowers are mostly in pairs; and the pods are strongly inflated (#211), ovoid to ellipsoidal, with an acuminate beak, and minutely white-hairy. The tip of the keel of the flower forms a prominent ascending beak.

211

211. Bladder Locoweed in fruit, near Mt. Edith Cavell, Alberta.

212

212. Bladder Locoweed, *Oxytrois podocarpa,* Iqaluit, Baffin Island. The leaves and flowering stems spread out from the center.

213

Hippuridaceae, Mare's-tail Family

213. Mare's-tail, *Hippuris vulgaris,* Churchill, Manitoba. The stems arising from rootstocks emerge above the water and bear linear leaves in close whorls. Widely distributed mainly south of the 70°N latitude, but also found in northern Ellesmere Island.

214

Lentibulariaceae, Bladderwort Family

214. Common Butterwort, *Pinguicula vulgaris,* Churchill, Manitoba. The yellowish geen leaves are all basal, inrolled at the margins, obtuse, and greasy or sticky. The purplish, nodding flowers are solitary at the end of the flowering stems. The corolla is 2-lipped and 5-lobed; the lower lip is longer than the upper one, 3-lobed, and prolonged into a spur at the bse. Subarctic, carnivorous plants in moist habitats.

Juncaceae, Rush Family #23, 24, 48, 49

The Rush Family is closely related to the Lily Family. The Rushes and Wood-rushes are grass-like plants with tiny lily-like flowers with 6 scale-like perianth segments (sepals and petals which are alike).

215

Liliaceae, Lily Family #215-216

The traditional Lily Family is divided into many families, and *Tofieldia* is placed in the Malanthiaceae which includes *Veratrum* (False Hellebore) and *Zygadenus* (Death Camas).

215. Northern False Asphodel, *Tofieldia coccinea,* Repulse Bay, N.W.T. The tufted plants with chiefly basal leaves have purplish flowering stems with one leaf near the middle. The small purplish flowers are closely crowded. An arctic and subarctic species.

216

216. Common False Asphodel, *Tofieldia pusilla,* Iqaluit, Baffin Island. This species may be distinguished from the above species by its often taller, slender, green flowering stems which are leafless or with one small leaf near the base and by its whitish flowers. The short, flat leaves are 2-ranked and enfold each other at the base. Atctic, subarctic, and alpine.

217

Orchidaceae, Orchid Family

217. Small White Lady's-slipper or Sparrow's-egg Lady's-slipper, *Cypripedium passerinum,* Churchill, Manitoba. The 3 petals are white and the 3 sepals are green and shorter than the petals. The pouch-like lip (a modified petal) with a rounded opening and purple spots inside resembles superficially a sparrow's egg. The upper sepal is concave and the 2 lower sepals are free or united and closely appressed to the underside of the lip. Subarctic and boreal. The stems and leaves are soft-hairy.

218

Papaveraceae, Poppy Family

218. Arctic Poppy, *Papaver radicatum,* Pangnirtung, Baffin Island. The Arctic Poppy is a very hardy plant and is among the small number of herbaceous flowering plants at high altitude and latitude.

219

Plumbaginaceae, Leadwort Family

219. Thrift, *Armeria maritima,* Sachs Harbour, Banks Island. The Thrifts are densely tufted plants with very narrow, grass-like linear, basal leaves and pink flower-heads at the ends of the leafless flowering stems. The flower-heads are subtended by scarious involucral bracts, the outermost ones prolonging down the stem and uniting to form a sheathing tube. The flowers have a conspicuous, tubular, ribbed calyx with a 5-lobed scarious plaited limb and a hairy tube. The short pink corolla is included in and free from the calyx. There are 5 stamens attached to the corolla tube and one superior ovary with 5 styles. Common on seashores, but also occur inland.

220

Poaceae (Gramineae), Grass Family
See also #63, 66, 119, 150

220. Alpine Foxtail, *Alopecurus alpinus,* Repulse Bay, N.W.T. This grass is usually found in wet or moist tundra. The inflorescence is compact, ovoid or oblong, and hairy. The spikelets are flattened, short-stalked, disariculating below the glumes, 1-flowered. The glumes are hairy, united at the base, enclosing the awned lemma. The awn is attached to the back below the middle. This grass often forms a dense turf near settlements where the soil is enriched by human refuse.

221

Polygonaceae, Buckwheat Family
See also #60

221. Viviparous Knotweed or Alpine Bistort, *Polygonum viviparum*, Churchill, Manitoba. The lower flowers are replaced by bulblets that often sprout on the parent plants and fall off to grow new plants. This perennial herb with a short thick rootstock is also often abundant near human habitations.

222

Primulaceae, Primrose Family #222-223

222. Rock-jasmine, *Androsace chamejasme*, Parker Ridge, Alberta. This is a circumpolar, arctic-alpine species common in northwestern America. The small hairy leaves are in rosettes which form clumps. The corolla has a constricted throat forming a yellow eye, and the lobes are entire. The corolla-tube is shorter than the calyx.

223

223. *Primula egaliksensis*, Churchill, Manitoba. This subarctic Primrose has deeply 2-cleft, narrow corolla-lobes; and entire-margined or obscurely undulate-dentate, slender-stalked leaves in basal rosettes.

224

225

Pyrolaceae, Wintergreen Family

224. Arctic or Large-flowered Wintergreen, *Pyrola grandiflora*, eastern Brooks Range, Alaska. The leathery leaves are shiny above. The flowers have spreading petals, a strongly declined style, upcurved filaments, and lemon-yellow anthers. An arctic and subarctic species.

226. Anemone drummondii
The Whistlers, Jasper P., Alb.
Tufted; leaves deeply divided.
Cordilleran.

226

Ranunculaceae, Buttercup Family
#225-232
See also #36, 99, 143

225-~~226~~. Northern Anemone, *Anemone parviflora*, Holman, Victoria Island ~~(225); Whistlers Mountain, Alberta (226)~~. The flowers (sepals) are white above, tinged with blue beneath. The leaves on the flowering stems are nearly stalkless and whorled. The basal leaves are deeply 3-parted into broadly wedge-shaped segments, these again lobed. The stems are more or less hairy. Low-arctic, subarctic, and alpine in distribution.

227

227. White Water-crowfoot, *Ranunculus aquatilis* var. *subrigidus* (*R. circinatus* var. *subrigidus, R. subrigidus, Batrachium circinatum* subsp. *subrigidum*), Churchill, Manitoba. The totally submersed plants have finely dissected, essentially stalkless leaves at the tips of dilated hairy stipules. Boreal and subarctic; also in some Canadian arctic islands.

228

228. Seaside Crowfoot or Creeping Buttercup, *Ranunculus cymbalaria,* Churchill, Manitoba. This Buttercup with creeping stems grows on wet ground near the seashore. The leaves are long-petioled, ovate-oblong, more or less cordade at the base, and shallowly round-toothed on the margins. A subarctic species.

229

229. Arctic or Floating Buttercup, *Ranunculus hyperboreus,* Skeleton Lake edge, near Lake Hazen, Ellesmere Island, 07-31-87. This small plant is creeping in moss on land. It also grows in the water in aquatic moss with floating stems and leaves (#143, 144). The leaves are deeply 3-lobed, with the lobes rounded at the apex. The small axillary flowers are about 5 mm across, with 3-5 petals, 3-5 sepals, and several stamens. An arctic and subarctic species, ranging south to Colorado.

230

231

230. Yellow Water-crowfoot, *Ranunculus gmelinii,* Churchill, Manitoba. The submersed leaves are deeply divided into 3-5 wedge-shaped lobes, each lobe again cleft into linear divisions. The aerial leaves are less finely divided. This is a more southern species than #229, being low-arctic and subarctic in distribution, extending south to Colorado.

231. Snow Buttercup, *Ranunculus nivalis,* Pond Inlet, Baffin Island. The basal leaves are heart to kidney-shaped at the base, deeply 3-lobed, and the lobes are cleft. The flowers are large, and the sepals are densely covered with long, dark reddish-brown hairs on the back. An arctic species often found in wet moss by brooks in snow-accumulation areas.

232

232. Dwarf Buttercup, *Ranunculus pygmaeus,* Sachs Harbour, Banks Island. These small tufted buttercups have small flowers. The weak flowering stems are ascending or spreading. The petals are about as long as the sepals. The basal leaves are truncate to subcordate at the base, deeply 3-lobed, with the lateral lobes 2- to 3-cleft or 2- to 3-toothed. An arctic species of late snow-melt areas predominated by mosses extending south to Colorado.

233

234

235

236

Rosaceae, Rose Family #233-239
See also #64, 70, 92, 93

233-234. Arctic Avens, *Dryas integrifolia,* Churchill, Manitoba. This mat-forming dwarf shrub is a common North American arctic, subarctic, and alpine species. The narrow leaves are dark green and shiny above, white-tomentose beneath, truncate or subcordate at the base; the margins are mostly turned under, entire or with few wavy teeth near the base. The plumes of the fruiting styles twist together when young, spreading apart later. For Dryas hummocks see #127, 153, 160, 161.

235. Arctic Cinquefoil, *Potentilla hyparctica,* Iqaluit, Baffin Island. The basal leaves with 3 leaflets are with rather thin whitish, straight hairs.

236. Red-stemmed Cinquefoil, *Potentilla rubricaulis,* Tanquary Fiord, Ellesmere Island. The dense basal leaves are pinnately divided into 5 leaflets and the undersides are densely covered by appressed, long, silky hairs, See also #70.

237

237. Alpine Avens, *Geum rosssii* (*Acomastylis rossii*), Trail Ridge, Colorado. These densely tufted or matted plants are common in the rocky alpine areas of the American Rockies. Also found on rock slopes in Eureka, Ellesmere Island (#93), Axel Heiberg and Melville islands of the Canadian Western Arctic, this species ranges westward to Alaska and eastern Asia. They also grow in moist sandy and gravelly places. The leaves are mostly basal, pinnately divided below and pinnately cleft above. The brown persistent stipules are nearly completely attached to the leaf-stalks. The stem leaves are greatly reduced. The leaves are hairy on the margins; stems are short-hairy especially towards the apex; the top-shaped calyx tube is moderately hairy; sepals and slightly shorter bractlets are sparingly hairy; the styles are glabrous; and the ovaries are densely hairy.

238

238. *Potentilla egedii*, Churchill, Manitoba. The plants grow gregariously in saline habitats along the low-arctic and subarctic seacoasts. They are spreading by slender runners, and have pinnately divided leaves and solitary flowers on long, leafless stalks. The leaflets are 5 to 7, not silvery pubescent beneath.

239

239. Sibbaldia, *Sibbaldia procumbens*, Mt. Goliath, Colorado. This alpine and subarctic, dwarf, matted, perennial herb has compound leaves of 3 leaflets toothed at the apex and very small flowers with narrow yellow petals shorter than the sepals and 5 stamens.

240

241

242

243

Salicaceae, Willow Family #240-247

240-242. Arctic Willow, *Salix arctica,* Holman, Victoria Island (240); Beartooth Mountains, Wyoming (241); Baffin Island (242). The prostrate plants have entire-margined leaves which are silky-hairy on the undersides when young. The hairy scales of the catkins are rounded at the apex, blackish above, and reddish-brown below. The ovaries are densely hairy; styles rather long; stigmas reddish-brown, bifid.

243. Net-veined Willow, *Salix reticulata,* Arctic Bay, Baffin Island. The male catkins of this creeping willow have reddish brown scales. See also #245.

244

244. Arctic Willow, *Salix arctica,* Repulse Bay, N.W.T. The hairy seed pods are borne on leafy stalks. The apex of most leaves is pointed. This arctic-alpine willow reaches the Far North (#131 , 134-137).

245

245. Net-veined Willow, *Salix, reticulata,* Churchill, Manitoba. The reddish seed pods with short whitish hairs are borne on leafless stalks at the end of the shoots. The entire-margined leaves are rounded at the apex, strongly net-veined, whitish on the underside, and relatively long-stalked. This arctic-alpine willow ranges from Alaska to Baffin Island, extending south to the Rockies, but does not occur in the Canadian Far North or Greenland. In Baffin Island it is frequently found in snowpatch communities.

246

246. Herb-like Willow, *Salix herbacea,* Pangnirtung, Baffin Island. These very low willows grow gregariously in snow-accumulation areas of the Canadian Eastern Arctic south of 75°N, Greenland, and Europe. They have slender, subherbaceous stems buried in moss; round leaves with slightly cordate base, and wavy-toothed margins; and few hariless pods on short stalks.

247

247A

247-247A. Herb-like Willow, *Salix herbacea,* Pangnirtung Fiord, Baffin Island. The male catkins have yellowish-green scales. The female catkins have reddish scales and short styles with bifid stigmas.

Saxifragaceae, Saxifrage Family, #248-255
See also #26, 30, 130, 132, 159

248. Yellow Mountain Saxifrage, *Saxifraga aizoides,* Repulse Bay, N.W.T. The matted plants have leafy stems with solitary to few flowers. The linear, fleshy leaves are sharp-pointed at the tip. Arctic-alpine species, absent from Alaska.

249. Spider Plant, *Saxifraga flagellaris,* Mt. Chamberlin, Brooks Range, Alaska. The plant produces radiating, whip-like, arching runners from the basal rosette of leaves, which form new tiny rooting rosettes at the tip. The basal leaves are acute at the apex, and bristly on the margins. The leafy stem bears a few yellow flowers with green, flat base of the calyx (subsp. *flagellaris*). This race occurs in Alaska and eastern Siberia. In eastern N. America (#26), the plants have mostly single flowers. The stem, stem-leaves, and sepals are dark purple, and beset with stalked glandular hairs with purple heads. The base of the calyx is top-shaped (subsp. *platysepala*). The Spider Plant also occurs in the Rockies south to Colorado (*Hirculus platysepalus* subsp. *crandallii*).

248

249

250

250. Tufted Saxifrage, *Saxifraga caespitosa*. Iqaluit (Frobisher Bay), Baffin Island. These tufted plants have closely crowded basal leaves which are broadly wedge-shaped and deeply 3-lobed, with linear-oblong, obtuse lobes. The stems with reduced leaves bear a few (3) white flowers. This (subsp. *eucaespitosa*) occurs chiefly in the Canadian Eastern Low Arctic and Subarctic. In the Canadian High Arctic (#30) and in Alaska, the flowers are mostly solitary with dark purple calyx (subsp. *uniflora*). The Tufted Saxifrage also occurs in the Rockies south to Colorado (*Muscaria monticola*, *Saxifraga caespitosa* subsp. *monticola*, *M. delicatula*).

251. Nodding or Bulblet Saxifrage, *Saxifraga cernua*, Repulse Bay, N.W.T. The plants grow somewhat gregariously as in this picture or solitarily. The stems bear red bulblets in the axils of reduced leaves below the terminal solitary flowers that nod when young. The basal and lower stem-leaves are kidney-shaped, palmately 5-7 lobed, and long-talked. They grow on dry rocky slopes as well as in wet places such as in wet moss by brooks in snow-accumulation areas. Circumpolar, arctic-alpine, south to Colorado.

252. Prickly Saxifrage or Three-toothed Saxifrage, *Saxifraga tricuspidata*, Churchill, Manitoba. The closely crowded basal leaves are narrowly wedgeshaped, with 3 prickly teeth at the tip, stiff, and often reddish tinged (# 129). The stems with few reduced leaves bear few flowers. The white petals are orange-dotted. The plants are densely matted (#129) in dry, rocky and gravelly places. North American arctic-alpine species, extending south to the Canadian Rockies.

251

252

253

254

255

253-255. Purple Saxifrage, *Saxifraga oppositifolia,* Iqaluit (253), Auyuittuq National Park (254), Baffin Island; Repulse Bay, N.W.T. (255). The Saxifrages have 10 stamens and 2 styles which spread in fruit and become the 2 beaks of a casule. The Purple Saxifrage has purple crimped petals, purple filaments and styles; crowded, prostrate branches covered with scale-like, overlapping, 4-ranked, opposite leaves which are hairy on the margins. It occurs chiefly in calcarious, gravelly places throughout the Arctic and in the alpine areas of the Rockies south to Wyoming.

256

Scrophlariaceae, Figwort Family #256-262
See also #58, 75-78

The Louseworts (*Pedicularis*) have pinnately divided leaves and 2-lipped corollas. The upper lip (galea) is laterally compressed into a helmet or hood shape, and encloses 4 stamens attached to the corolla. The lower lip is 3-lobed. The galea is neither beaked nor toothed near the apex in #256, 259, 58, 77: beaked in #260; not beaked but with teeth near the apex in #258, 76.

256-257. Capitate Lousewort, *Pedicularis capitata,* Lake Schrader, Brooks Range, Alaska (256); Repulse Bay, N.W.T (257). The few large flowers are in a terminal compact cluster subtended by leafy bracts. The corolla is yellowish-white, somewhat tinged with purple at the tip. The stem is nearly leafless. This more or less short hairy, arctic-alpine species extends southward to the Canadian Rockies.

258, 261-262. Sudetan Lousewort, *Pedicularis sudetica,* Thoro, Alaska Range, Alaska (258, 261); Repulse Bay, N.W.T. (262). These alpine and arctic plants grow in rather wet places, extending south to British Clumbia, but absent from Greenland. The large corolla is generally dark reddish-purple; the galea is strongly arched, with 2 prominent teeth near the tip. The calyx and bracts are densely woolly. The stem is nearly leafless, hairless, and purplish above.

257

258

259

260

259. Flame-tipped Lousewort, *Pdeicularis flammea,* Iqaluit, Baffin Island. These plants with reddish-purple, leafy, simple (unbranched) stems are essentially hairless. The lower lip of the corolla is yellow, and the galea is dark purple at the tip, neither beaked nor toothed. They grow in moist places. Arctic and subarctic.

260. Lapland Lousewort, *Pedicularis lapponica,* Iqaluit, Baffin Island. The pale yellow, fragrant flowers are in short clusters. The uper lip of the corolla has a rather broad beak. The purplish-brown stems are leafy. This low-arctic and subarctic species has a slender creeping rootstock, and often grows gregariously in moist places.

261-262. See #258.

261

262

Although tree-sized willows rarely occur in the Acrtic, coniferous trees do not grow there. This is the tundra. Between the tundra and the predominantly coniferous forest to the south, there is a transitional zone in which most of the land is still occupied by tundra vegetation and stands of conifers are only in especially favorable locations, such as shorelines of lakes, ponds, rivers, and rivulets due to deeper thawing of the surface layer for proper functioning of tree roots; and sheltered valleys due to protection from desiccating winds during the long winter when trees grow on frozen ground. This is the forest-tundra zone and is the Subarctic proper. Northern Labrador, approximately between latitudes 58° and 55°N, is covered by the forest-tundra. Also usually included in the subarctic zone is the northern part of the coniferous forest, where generally low trees (mostly spruces) are widely spaced and the open spaces between them are covered with a carpet of lichens (#263-269). This is the lichen woodland. In eastern Canada, the southern boundary of the lichen woodland is roughly the 52nd parallel of north latitude through southwestern Labrador and central Quebec. In western Canada, the boundary line skirts Southern Indian Lake in Manitoba, Great Slave and Great Bear lakes in Northwest Territories. In Yukon the land around the Arctic Circle is subarctic.

In the subarctic region, the growing season is longer, summer temperatures and annual precipitation are higher than in the arctic region. In southern Labrador, winter comes in Ocotober with its snow cover and ends in May when the snow begins to disappear. In summer the mean daily maximum temperatures exceed 18°C (65°F) for nearly a month. The mean total annual precipitation is 700-1200mm(27-47 inches), about half of it in the form of rain and half as snow.

263

263. Reindeer (Caribou) Lichen, *Cladonia stellaris* (*C. alpestris, Cladina alpestris*), southeastern Labrador. This lichen is common in Labrador and Alaska and is eaten by caribou. It is said that the recovery of a grazed Reindeer Lichen cover takes 10-20 years, the growth rate being only about 2 mm per year. When burned, the full recovery is said to take longer.

264

264. Reindeer (Caribou) Lichen, *Cladonia rangiferina* (*Cladina rangiferina*), southeastern labrador. This lichen is not so abundant as *Cladonia stellaris* in Labrador and Alaska, but is one of the common Reindeer Lichens of the arctic and subarctic regions.

265

265. Curled Lichen, *Cetraria cucullata*, Anaktuvuk Pass, Alaska. Lichens dry easily and absorb water rather quickly from rain, fog or the atmosphere. They carry on photosynthesis most effectively during and after periods of fog and rain showers. On days when precipitation does not occur, the peak of photosynthesis takes place in the very early morning hours (2:00 to 8:00) when the photosynthetic tips are wettest. Lichens are more luxuriant in the relatively moister Low Arctic and Subarctic than in the drier arctic regions.

266

266. Snow Lichen, *Cetraria nivalis*, Pond Inlet, Baffin Island. The tips of flattened lobes are finely divided and crinkled. This is an arctic, alpine, and boreal species. The gray, horn-shaped lichen mixed in the picture is Worm Lichen (#270).

267

268

269

270

267. *Stereocaulon alpinum,* Anaktuvuk Pass, Alaska. This lichen is common in uplands of tundra with dwarf shrubs (#194).

268. *Stereocaulon pashale,* Eagle Plain, Dempster Highway, Yukon. Common in the northern Black Spruce woods of the western N. American Subarctic.

269. *Nephroma arctica,* Anaktuvuk Pass, Alaska. It is said that this is a lichen caribou does not eat.

270. Worm Lichens, *Thamnolia vermicularis* and its chemical strain *T. subuliformis,* Denali National Park, Alaska. These lichens are circumoplar, arctic-alpine in distribution.

271

271. *Coelocaulon divergens,* (*Alectoria divergens*) Pond Inlet, Baffin Island. This and the following species as well as the Snow Lichen (#266) commonly grow in the rather dry tundra (#2, 88).

272

272. Mane Lichen, *Alectoria ochroleuca,* Anaktuvuk Pass, Alaska, This is an intricately branched lichen with black, hair-like tips and yellowish or brownish-yellow lower parts. A circumpolar, arctic-alpine species.

273

273. Brown Finger Lichen, *Dactylina arctica,* Anaktuvuk Pass, Alaska. These finger-like or horn-shaped lichens are often found in late snow melt areas. This is also a circumpolar, arctic-alpine species.

274

274 and 277. Coastal forest-tundra, near Capstan Island on the southeastern coast of Labrador, 08-21-82. The southern Atlantic coast of Labrador, under the influence of the cold Labrador current and arctic air masses, develops the forest-tundra zone. The scrubby trees are Black Spruce (*Picea mariana*) and Balsam Fir (*Abies balsamea*). Annual precipitation of this region is approximately 1000 mm (40 inches).

275

276

275-276. Balsam Fir, *Abies balsamea*, in #274 and 277. The balsam fir occurs from Labrador coast southward to the northern Appalachians and westward to western Alberta. In the typical form, the bracts on the cone scales are small and concealed by the scales in intact cones. In the bracted variety (var. *phanerolepis*), the bracts are slightly longer than their scales, protruding beyond them as shown in the picture. This variety occurs in Labrador, Newfoundland, along the coast of New Brunswick and Maine, and at higher altitudes in the mountains of Maine and New Hampshire.

277

278

278. Lichen-heath carpet in #274 and 277. Among the dense extensive growth of Reindeer Lichen (*Cladonia stellaris*) are Alpine Bilberry, Black Crowberry, Labrador-tea, Mountain Cranberry, and Baked-apple Berry (#279).

279

279. Baked-apple Berry or Cloudberry, *Rubus chamaemorus*, in #278. These ripe berry-like fruits are very good to eat. Only female plants bear fruits which ae red when immature. The flowers are white.

280

281

280-281. Fores-tundra, near Capstan Island, southeastern Labrador. The hill top is occupied by a community dominated by dwarf heath shrubs and lichens. The chief species are Alpine Bilberry, Black Crowberry, Mountain Cranberry, Reindeer Lichens and Snow Lichen. Also some Labrador-tea and Baked-apple Berry grow here.

282

282. Alpine Bilberry (*Vaccinium uliginosum* var. *alpinum*) among lichens (*Cladonia* sp.), southeastern Labrador.

283

283. Krummholz (scrub forest) of Balsam Fir on hilltop, near Capstan Island, southeastern Labrador. In the exposed, dry, gravelly place Pink Crowberry (# 284) grows.

284

285

284-285. Pink Crowberry, *Empetrum eamesii* (*E. rubrum* subsp. *eamesii*), in #283. The Pink Crowberry is lower and more compact than the Black Crowberry (#192) and prefers dry exposed habitats. Young twigs are densely white-hairy, while they are glabrous in the latter. It ranges from southeastern Labrador to Newfoundland and Nova Scotia.

Selected References

Andrews, J.T., W.N. Mode, P.J. Webber, G.H. Miller, and J.D. Jacobs. Report on the distribution of dwarf birches and present pollen rain, Baffin Island, N.W.T. Arctic 33: 50-58, 1980.

Babb, T.A. and Bliss, L.C. Susceptibility to environmental impact in the Queen Elizabeth Islands. Arctic 27: 234-237, 1974.

Barrett, P.E. and Teere, J.A. Vascular plants of the Truelove Inlet region, Devon Island. Arctic 26: 58-67, 1973.

Bell, K.L. and Bliss, L.C. Overwinter phenology of plants in a polar semidesert. Arctic 30: 118-121, 1977.

Beschel, R.E. Botany; and some remarks on the history of vegetation and glacierization. *In* Jacobsen-McGill Arctic Research Expedition to Axel Heiberg Island, Queen Elizabeth Islands. Preliminary Report 1959-1960, p. 179-200, 1961. McGill University, Montreal.

————. Geobotanical studies on Axel Heiberg Island in 1962. *In* Müller, F. et al. Axel Heiberg Island Research Reports, Preliminary Report 1961-1962, p. 199-215, 1963. McGill University, Montreal.

Bird, J.B. The physiography of arctic Canada, with special reference to the area south of Parry Channel. 1967. John Hopkins Press, Baltimore.

Bliss, L.C. Arctic heathlands. *In* Specht, R.L. (ed) Heathlands and related shrublands: Descriptive studies. p. 415-424, 1979.

Böcher, T.W. Eco~~logical~~geographical classification of arctic vegetation based on shoots density determinations. Meddelelser om Grønland 199 (6): 1-61, 1977.

Böcher, T.W., K. Holmen, and K. Jakobsen. The flora of Greenland. 1968. English translation from the Danish edition Grønlands flora, 2, revised edition 1966. P. Hasse & Sons, Copenhagen.

Brassard, G.R. The plant habitats of the Tanquary Camp area, Ellesmere Island, N.W.T. Defence Research Board, Canada: Geophysics, Hazen 32, 1968.

Brassard, G.R. and Beschel, R.E. The vascular flora of Tanquary Fiord, northern Ellesmere Island, N.W.T. Canadian Field-Naturalist 82: 103-113, 1968.

Brassard, G.R. and Longton, R.E. The flora and vegetation of Van Hauen Pass, northwestern Ellesmere Island. Canadian Field-Naturalist 84: 357-364, 1970.

Bruggemann, P.F. and Calder, J.A. Botanical investigations in northeast Ellesmere Island, 1951. Canadian Field-Naturalist 67: 157-174, 1953.

Edlund, S.A. Vegetation of Lougheed Island, District of Franklin. *In* Current Research, Part A, Geological Survey of Canada, Paper 80-1A, p. 329-333, 1980.

————. Bioclimatic zonation in a High Arctic region: central Queen Elizabeth Islands. *In* Current Research, Part A, Geological Survey of Canada, Paper 83-1A, p. 381-390, 1983.

England, J., L. Kershaw, C. Lafarge-England, and J. Bednarsaki. Northern Ellesmere Island: a natural resource inventory; submitted to Park Canada. 1981. Department of Geography, University of Alberta, Edmonton.

Environment Canada. Canadian Climate Normals, Temperature and Precipitaion 1951-1980, The North-Y.T. and N.W.T. 1982.

Fredskild, B. Contributions to the flora of Peary Land, North Greenland. Meddelelser om Grønland 178 (2): 1-23, 1966.

Holmen, K. The vascular plants of Peary Land, North Greenland. Meddelelser om Grønland 124 (9): 1-149, 1957.

Kuc, M. Vascular plants from some localities in the western and northern parts of the Canadian Arctic Archipelago. Canadian Journal of Botany 48: 1931-1938, 1970.

Maxwell, J.B. The climate of the Canadian Arctic Islands and adjacent waters. vol. 1. Climatological Studies 30, 1980. Atomospheric Environment Service, Toronto.

————. Climatic regions of the Canadian Arctic Islands. Arctic 34: 225-240, 1981.

Muc, M. and Bliss, L.C. Plant communities of Truelove Lowland. In Bliss, L.C. (ed), Truelove Lowland, Devon Island, Canada: a high arctic ecosystem. p. 143-154, 1977. University of Alberta Press, Edmonton.

Oliver, D.R. and Corbet, P.S. Aquatic habitats in a high arctic locality: The Hazen Camp study area, Ellesmere Island, N.W.T. Defence Research Board, Canada: Directorate of Physical Research, Geophysics, Hazen 26, 1966.

Packer, J.G. Polyploidy and latitude in the Canadian Arctic Archipelago. Arctic and Alpine Research 1: 15-27, 1969.

Polunin, N. Botany of the Canadian Eastern Arctic. III. Vegetation and ecology. National Museum of Canada, Bulletin 104, 1948.

————. The real Arctic: Suggestions for its delimitation, subdivision and characterization. Journal of Ecology 39: 308-315, 1951.

————. Circumpolar arctic flora. 1959. Clarendon Press, Oxford.

———. Introduction to plant geography. p. 380-422, 1960. McGraw-Hill Book Co., New York.
Porsild, A.E. Illustrated flora of the Canadian Arctic Archipelago. 1964. National Museum of Canada, Bulletin 146, 2nd ed.
Porsild, A.E. and Cody, W.J. Vascular plants of continental Northwest Territories, Canada. 1980. National Museums of Canada, Ottawa.
Powell, J.M. The vegetation and micro-climate of the Lake Hazen area, northern Ellesmere Island, N.W.T. Defence Research Board, Canada: Directorate of Physical Research, Geophysics, Hazen 14, 1961.
Rousseau, J. The vegetation of the Quebec-Labrador peninsula between 55° and 60°N. Naturaliste Canadien 9: 469-563, 1968.
Savile, D.B.O. The botany of Somerset Island, District of Franklin. Canadian Journal of Botany 37: 959-1022, 1959.
———. The botany of the northwestern Queen Elizabeth Islands. Canadian Journal of Botany 39: 909-942, 1961.
———. General ecology and vascular plants of the Lake Hazen Camp area. Arctic 17: 237-258, 1964.
Schofield, W.B. and Cody, W.J. Botanical investigations of coastal southern Cornwallis Island, Franklin District, N.W.T. Canadian Field-Naturalist 69: 116-128, 1955.
Thompson, D.C.A. A classification of the vegetaion of Boothia Peninsula and the northern District of Keewatin, N.W.T. Arctic 33: 73-99, 1980.
Young, S.B. The vascular flora of St. Lawrence Island, with special reference to floristic zonation in the arctic regions. Contributions from the Gray Herbarium of Harvard University, No. 201, p. 11-115, 1971.

Dorn, R.D. Manual of the vascular plants of Wyoming, 1977. 1498 p. in 2 vols. Garland Publishing, New York.
———. Vascular plants of Montana. 1984. Mountain West Publishing, Cheyenne, Wyoming.
Gleason, H.A. and Cronquist, A. Manual of vascular plants of northeastern United States and adjacent Canada. 1963. D. Van Nostrand Co., Princeton, New Jersey.
Moss, E.H. Flora of Alberta, 1983. 2nd ed. Revised by J.G. Packer. University of Toronto Press, Toronto.
Porsild, A.E. Rocky Mountain Wildflowers. 1974. Natural Series No. 2, National Museum of Canada and Parks Canada, Ottawa.
Scotter, G.W. and Flygare, H. Wildflowers of the Canadian Rockies. 1986. Hurting Publishers, Edmonton.
Simmons, H. and Miller, S. Notes on the vascular plant of the Mackenzie Mountain Barrens and surrounding area. Report No. 3, Northwest Territories Renewable Resources, Yellowknife, 1982.
Weber, W.A. Colorado flora: Western slope. 1987. Colorado Associated University Press, Boulder.

Cetraria tilesii, a calciphilous (lime-loving), arctic species. Anaktuvuk Pass, Brooks Range, Alaska, 07-29-79.

Index

Numerals are Photograph Numbers unless otherwise indicated

Abies balsamea 275, 276
Acomastylis rossii 93, 237
Alectoria divergens 271
ochroleuca 272
alkaligrass, creeping 116, 176
Alopecurus alpinus 220
alpine bistort 221
cress 25
foxtail 220
Alsinaceae p. 69
Androsace chamejasme 222
anemone, northern 225, 226
Anemone parviflora 225, 226
Arctagrostis latifolia 150
arctic, high p. 9; 13-18, 34
low 2
middle 86-88
arctic bladder pod 62
daisy 169
Arctoparmelia separata 52, 53
Arctostaphylos alpina 195
rubra 199
Arctous alpina 195
rubra 199
Arenaria peploides 174
rubella 180, 181
Armeria maritima 219
Arnica alpina 166
Asteraceae p. 67
Astragalus alpinus 206
avens, alpine 93, 237
arctic 127, 128, 134-137, 151, 153, 233, 234
azalea, alpine 202
baked-apple berry 279
Batrachium circinatum ssp. subrigidum 227
bearberry, alpine 194-196
red-fruited 199
Betula glandulosa 12
bilberry, alpine 84, 85, 203, 282
birch, dwarf 12
bird stone site 31
blueberry, arctic 203
Boraginaceae p. 69
Brassicaceae p. 72
Braya humilis ssp. arctica 156
purpurascens 154
thorild-wulffii 19, 20
buttercup, arctic 143, 229
birdfoot 99, 100
creeping 228
dwarf 232
floating 118, 143, 144, 229
northern 99, 100
snow 231
sulphur-colored 15, 36
butterwort, common 214
campion, moss 44
nodding bladder 40
Cardamine bellidifolia 25
Carex aquatilis var. stans 35
bigelowii 185
maritima 186
nardina 55, 56
stans 35
subspathacea 187
ursina 114, 117
Caryophyllaceae p. 69
Cassiope hypnoides 8, 9
tetragona 82, 83, 86, 87, 135, 158, 198, 200
Cerastium alpinum 104
regelii 125
Cetraria cucullata 265
nivalis 266
Chamaenerion latifolium 80
chamomile, seashore 170
chickweed, alpine 104
low 175, 176
polar 125
Regel's 125
chrysanthemum, entire-leaved 155
Chrysanthemum arcticum 169
integrifolium 155
cinquefoil, arctic 235
bright 64
red-stemmed 70, 236
Vahl's 91, 92
Cladina *see* Cladonia
Cladonia alpestris 263
rangiferina 264
sp. 282
stellaris 263, 278
cloudberry 279
club-moss, fir 165, mountain 165
Cochlearis officinalis 27, 61, 182
Coelocaulon divergens 271
Compositae p. 67
cotton-grass 37
cotton-grass, narrow-leaved 188
Scheuchzer's 189
sheathed 190
tall 188
cranberry, mountain 204, 205
Crepis nana 167
crowberry, black 192, 193
pink 284, 285
crowfoot, seaside 228
Cruciferae p. 72
Cyperaceae p. 73

Cypripedium passerinum 217
Cystopteris fragilis 45, 46
Dactylina arctica 54A, 273
Diapensia lapponica 191
Diapensiaceae p. 75
Draba bellii 50, 51
corymbosa 50, 51
macrocarpa 50, 51
Dryas integrifolia 71, 72, 127, 135, 151, 152, 233, 234
Dryopteris fragrans 163
Dupontia fisheri 119
Empetraceae p. 75
Empetrum eamesii 284, 285
nigrum 192, 193
rubrum ssp. eamesii 284, 285
Epilobium latifolium 80
Equisetum variegatum 147
Ericaceae p. 76
Erigeron compositus 69
eriocephalus 97, 98
humilis 168
unalaschkensis 168
Eriophorum angustifolium, 188
ssp. triste 37
scheuchzeri 65, 189
spissum 190
triste 37
vaginatum ssp. spissum 190
Erysimum pallasii 183, 184
Eutrema edwardsii 39
Fabaceae p. 80
false asphodel, common 216
northern 215
Family, Bladderwort, p. 82
Borage p. 69
Buckwheat p. 85
Buttercup p. 86
Chickweed p. 69
Crowberry p. 75
Diapensia p. 75
Fern p. 66
Figwort p. 96
Grass p. 84
Heath 2, 4, 5, p. 76
Leadwort p.84
Lily p. 83
Mare's-tail p. 82
Mustard p. 72
Orchid p. 83
Pea 2, 6, p. 80
Pink p. 69
Poppy p. 84
Primrose p. 85
Rose p. 89
Rush p. 82
Saxifrage p. 93, Sedge p. 73
Sunflower p. 67
Willow p. 91
Wintergreen p. 86
fern, brittle 45, 46
fragile 45, 46
fragrant 163
smooth cliff 47
Festuca baffinensis 63
fir, balsam 275, 276, 281.
fir club-moss 165
fireweed, dwarf 80
fleabane, arctic 97, 98
cutleaf 69
mountain 168
forest-tundra p. 98; 274, 277, 280
Geum rossii 93, 237
Gramineae p. 84
Harrimanella hypnoides 8, 9
hawksbeard, dwarf 167
heather, arctic 54, 82, 83, 86, 87, 135, 158, 197, 198, 200
moss 7, 8, 9
mountain 10, 11
Hippuridaceae p. 82
Hippuris vulgaris 142, 213
Honckenya peploides 174
horsetail 147
hummocks 71, 73, 82, 83, 127, 128, 134, 135, 151, 153, 160, 161
Juncaceae p. 82
Juncus biglumis 23, 24
knotweed, viviparous 221
Kobresia myosuroides 72, 154
Labrador-tea 201
Labrador-tea, narrow-leaved 4
northern 4
lady's-slipper, small white 217
sparrow's-egg 217
Ledum decumbens 4
groenlandicum 201
palustre ssp. decumbens 4
Leguminosae 2, 6, p. 80
lemming den site 31
Lentibulariaceae p. 82
Lesquerella arctica 62
lichen 2, 88, 280-282
lichen, brown finger 54A, 273
curled 265
mane 272
reindeer 263, 264, 278
snow 266
lichen woodland p. 98
Liliaceae p. 83
locoweed, bladder 210-212
Loiseleuria procumbens 202
lousewort, arctic 75, 76
capitate 256, 257
flame-tipped 259
hairy 77, 78
Lapland 260
Sudetan 258, 261, 262

woolly 58, 78A
Luzula arctica 48
confusa 49
nivalis 48
Lycopodium selago 165
mare's-tail 142, 213
marsh-fleabane 171
mastodon flower 171
Matricaria ambigua 170
Melandrium apetalum 40
triflorum 18, 75
Mertensia maritima 172
milk-vetch, alpine 206
Minuartia rubella 180, 181
mountain sorrel 60
Nephroma arctica 269
Orchidaceae p. 83
Oxyria dignya 60
Oxytrope, arctic 208
inflated 210
Maydell's 6
silvery 209
Oxytropis arctica 208
arctobia 209
maydelliana 6
podocarpa 210-212
oysterleaf 172
Papaver radicatum 94, 95, 218
Papaveraceae p. 84
Parmelia separata 52, 53
Parrya arctica 32, 33
pearlwort, tufted 177-179
Pedicularis arctica 75, 76
capitata 256, 257
flammea 259
hirsuta 77, 78
lanata 58, 78A
langsdorfii ssp. arctica 75, 76
lapponica 260
sudetica 258, 261, 262
Phyllodoce caerulea 10, 11
Pinguicula vulgaris 214
Pleuropogon sabinei 66
Plumbaginaceae p. 84
Poaceae p. 84
polygon 28, 71, 160
Polygonaceae p. 85
Polygonum viviparum 221
poppy, arctic 94, 95, 218
Potentilla egedii 238
hyparctica 235
pulchella 64
rubricaulis 70, 236
vahliana 91, 92
Primula egaliksensis 223
Primulaceae p. 85
Puccinellia phryganodes 116, 176
Pyrola grandiflora 224
Pyrolaceae p. 86
regwort, marsh 171
Ranunculaceae p. 86
Ranunculus aquatilis var. subrigidus 227
circinatus var. subrigidus 227
cymbalaria 228
gmelinii 230
hyperboreus 118, 142-144 229
nivalis 231
pedatifidus var. leiocarpus 99, 100
pygmaeus 232
sabinei 36
subrigidus 227
sulphureus 15, 36
Rhacomitrium lanuginosum 54
Rhododendron lapponicum 5
river beauty 80
rock-jasmine 222
Rosaceae p. 89
Rosebay, Lapland 5
Rubus chamaemorus 279
rush, two-flowered 23, 24
Sagina caespitosa 177-179
Salicaceae p. 91
Salix arctica 131, 240-242, 244
herbacea 246, 247
reticulata 243, 245
sandwort, reddish 180, 181
seabeach 173, 174
Saxifraga aizoides 248
caespitosa 30, 250
cernua 251
flagellaris 26, 249
hirculus 38, 38A
nivalis 130
var. tenuis 132
oppositifolia 253-255
rivularis 159
tenuis 132
tricuspidata 129, 252
Saxifragaceae p. 93
Saxifrage, alpine 130
brook 159
bulblet 251
nodding 251
prickly 129, 252
purple 253-255
tree-toothed 252
tufted 30, 250
yellow marsh 38, 38A
yellow mountain 248
Scrophlariaceae p. 96
scurvy-grass 27, 61, 182
sea lungwort 172
sea-purslane 174
Senecio congestus 171
sedge, bear 114, 117

Bigelow's 185
curved 186
Hoppner 187
Nard 55, 56
water 35, 145, 146
Sibbaldia procumbens 239
Silene acaulis 44
sorensensis 18
uralensis 40
spider plant 26, 249
Stellaria humifusa 175
Stereocaulon alpinum 267
pashale 268
subarctic p. 98
Taraxacum phymatocarpum 101
pumilum 21
thrift 219
Tofieldia coccinea 215
pusilla 216
Vaccinium gaultheroides 203
uliginosum var. alpinum 84, 85, 203, 282
vitis-idaea var. minus 204
vegetation, continuous 2, 3, 34, 87, 88
sparse 16, 18
wallflower, Pallas's 183, 184
water-crowfoot, white 227
yellow 230
willow, arctic 131, 134-137, 240-242, 244
herb-like 246, 247
net-veined 243, 245
willow-herb, broad-leaved 80
wintergreen, arctic 224
large-flowered 224
wood-rush, arctic 48
northern 49
snow 48
Woodsia glabella 47, 138, 139

Appendix. Vascular plants common to the northernmost parts of Ellesmere Island and Greenland north of the 81st parallel.

The Ellesmere region north of the line joining Greely Fiord and Archer Fiord, called Grant Land, includes the areas of Van Hauen Pass (81°07′N, 85°55′W), Tanquary Camp (81°25′N, 76°55′W), Hazen Camp (81°49′N, 71°18′W), and Alert (82°30′N, 62°20′W). The Greenland region includes Peary Land, the area north of Independence Fiord (approximately 82°N) and east of Victoria Fiord (approximately 48°W). Grant Land supports essentially the whole flora of Peary Land and some additional species. Approximately 100 species are common to the two regions. The numerals in the following list denote illustrations in this book.

Agropyron latiglume (Roegneria borealis)
Alopecurus alpinus 220
Arctagrostis latifolia 150
Arenaria (Minuartia) rossii
rubella 180, 181
Armeria maritima (A. scabra) 219
Braya humilis ssp. arctica (Torularia humilis) 156
purpurascens 154
thorild-wulffii 19, 20
Calamagrostis purpurascens
Cardamine bellidifolia 25
Carex aquatilis var. stans (C. stans) 35
maritima 186
misandra
nardina 55, 56
rupestris
saxatilis
Cassiope tetragona 200
Cerastium alpinum 104
arcticum
regelii 125
Cochlearia officinalis (C. groenlandica) 27, 61
Colpodium vahlianum
Cystopteris fragilis 45, 46
Deschampsia brevifolia
Draba alpina
cinerea
corymbosa (D. bellii) 50, 51
lactea
nivalis
oblongata
subcapitata
Dryas integrifolia 233, 234
Epilobium (Chamaenerion) latifolium 80
Equisetum arvense
variegatum 147
Erigeron compositus 69
eriocephalus 97, 98
Eriophorum scheuchzeri 189
triste 37
Erysimum pallasii 183, 184
Eutrema edwardsii 39

Festuca baffinensis 63
brachyphylla
hyperborea
Hierochloe alpina
Hippuris vulgaris 213
Juncus albescens (J. triglumis)
biglumis 23, 24
castaneus
Kobresia myosuroides
Lesquerella arctica 62
Luzula confusa 49
nivalis (L. arctica) 48
Lycopodium selago 165
Melandrium affine
apetalum ssp. arcticum 40
triflorum 18, 75
Oxyria digyna 60
Papaver radicatum 95, 218
Pedicularis capitata 256, 257
hirsuta 77. 78
Phippsia algida
Pleuropogon sabinei 66
Poa abbreviata
alpigena var. colpodea
arctica ssp. caespitans
glauca
hartzii
Polygonum viviparum 221
Potentilla hyparctica 235
nivea ssp. chamissonis (P. chamissonis)

The author at Hazen Camp, 1987.

pulchella 64
rubricaulis 70, 236
vahliana 91, 92
Puccinellia angustata
phryganodes 116, 176
Ranunculus aquatilis var. eradicatus (R. trichophyllus var. eradicatus, R. confervoides)
hyperboreus 143, 144, 229
pedatifidus var. leiocarpus (R. affinis) 99, 100
sabinei
sulphureus 15, 36
Sagina intermedia
Salix arctica 240-242, 244
Saxifraga caespitosa 30
cernua 251
flagellaris ssp. platysepala 26
foliolosa
hyperborea (S. rivularis ?)
nivalis 130
oppositifolia 253-255
rivularis 159
tenuis 132
tricuspidata 129, 252
Silene acaulis 44
Stellaria crassipes
edwardsii
Taraxacum arcticum
arctogenum
hyparcticum
phymatocarpum 101
pumilum 21
Trisetum spicatum
Vaccinium uliginosum var, alpinum (V. uliginosum ssp. microphyllum) 85, 203, 282
Woodsia glabella 47, 138, 139